OECD Urban Studies



# The Circular Water Economy in Latin America

OECD

BETTER POLICIES FOR BETTER LIVES

**Please cite this publication as:**
OECD (2025), *The Circular Water Economy in Latin America*, OECD Urban Studies, OECD Publishing, Paris, https://doi.org/10.1787/a0508572-en.

ISBN 978-92-64-89553-9 (print)
ISBN 978-92-64-88266-9 (PDF)
ISBN 978-92-64-60644-9 (HTML)

OECD Urban Studies
ISSN 2707-3432 (print)
ISSN 2707-3440 (online)

**Photo credits:** Cover © Abstract Aerial Art/Getty Images Plus.

Corrigenda to OECD publications may be found at: https://www.oecd.org/en/publications/support/corrigenda.html.

# Foreword

More frequent and severe climate shocks, coupled with demographic growth and rapid urbanisation, are putting increasing pressure on existing water infrastructure and services. Addressing these growing challenges requires bold, transformative action from both public and private sectors. This is especially the case for Latin America, which despite having the highest share of natural renewable water resources globally is particularly vulnerable to water risks. Without urgent and co-ordinated action, the risks of water scarcity and pollution are poised to escalate, jeapordising the health and safety of communities as well as the economic viability of businesses that depend on water resources.

A circular approach to the water economy offers a promising solution to these challenges. By prioritising water reuse, recycling, and the recovery of energy and materials from wastewater treatment, circular water models can foster more efficient water use, minimise water loss, and enhance resilience to climate change, while providing business opportunities in the water industry. Technological advancements and digitalisation are making these models increasingly viable, generating new opportunities for innovation in water management.

Governments across Latin America are beginning to recognise the importance of transitioning to the circular water economy. However, significant barriers to scaling up these efforts persist, including inadequate regulatory frameworks, insufficient financial resources, a lack of awareness among key stakeholders, and conflicting uses of water, with impacts on access to water for households, agriculture and industry.

Building on an OECD/Inter-American Development Bank (IDB) Survey on Water and Circular Economy, this report offers new insights on the state of the circular water economy in Latin America. It also proposes a roadmap for accelerating the transition across the region, with concrete actions for strengthening policy frameworks, adapting financing tools, and engaging stakeholders to shape a more circular water economy. The report ultimately calls for a systems approach to water management, which mainstreams circular water principles across various sectors of the economy.

Moving forward, joined-up action is necessary to unlock the full potential of the circular water economy. This report aims to guide policymakers, water operators, and businesses as they navigate towards a more sustainable and resilient water future for Latin America and beyond.

# Acknowledgements

This report was prepared by the OECD Centre for Entrepreneurship, SMEs, Regions and Cities (CFE) led by Lamia Kamal-Chaoui, Director, as part of the Programme of Work and Budget of the OECD Regional Development Policy Committee (RDPC) and the OECD Water Governance Programme.

The report was drafted by Oriana Romano, Head of Unit, and María Ferrer, Policy Analyst, Water Governance, Blue and Circular Economy Unit, under the supervision of Aziza Akhmouch, Head of the Cities, Urban Policies and Sustainable Development Division in the CFE. Maria Salvetti, Senior Water Economist, Florence School of Economics, provided substantial inputs and recommendations, contributing to earlier versions of the report. Georges Laimé, Policy Analyst, and Doros Voulgaridis, Intern, of the Water Governance, Blue and Circular Economy Unit, carried out data analysis and literature review. The report benefitted from comments by Nadim Ahmad, Deputy Director in the CFE, Anthony Cox and Soo-Jin Kim, Deputy Head of the Cities, Urban Policies and Sustainable Development Division in the CFE.

Warm thanks are conveyed to the IDB team for their invaluable contributions and comments throughout the policy dialogue and drafting process, in particular: Sergio Campos, Chief, Water and Sanitation Division; Maria Eugenia de la Pena, Water and Sanitation Lead Specialist, and Tania Paez Rubio, Water and Sanitation Senior Specialist.

Special thanks are also conveyed to the ten countries that responded to the OECD-IDB survey carried out between October 2023 and February 2024 and provided comments on the draft report (see the full list of respondents in Annex A).

Authors are grateful to Anthony Cox, Senior Policy Advisor, Ecologic Economics and former Deputy Director of the OECD Environment Directorate (ENV) for his in-depth review of the report.

Preliminary findings were presented at the 2nd IDB Workshop of the Regional Process of the Americas and the OECD-IDB Workshop "Circular Water Economy in selected LAC countries", both held in February 2024. Members of the OECD Water Governance Initiative (WGI) peer-reviewed the draft report during the 19th meeting (4-5 March 2024, Marrakesh, Morocco). WGI members Belén Ramos Alcalde, Head of International Affairs, Spanish Association of Water Supply and Sanitation (AEAS), and Juan Antonio Guijarro, Director of Institutional Relations, Public Policy and Alliance Development, Veolia Iberia Latam and Member of the Executive Committee, Aquae Foundation, participated in *ad hoc* interviews that helped improve the policy recommendations.

Thanks are extended to Jack Waters in the CFE for preparing the report for publication.

# Table of contents



## FIGURES

## INFOGRAPHICS



## TABLES



## BOXES

# Abbreviations and acronyms

| | |
|---|---|
| **ECLAC** | Economic Commission for Latin America and the Caribbean |
| **EU** | European Union |
| **EUR** | Euro |
| **FAO** | Food and Agriculture Organization |
| **GHG** | Greenhouse Gas |
| **GDP** | Gross Domestic Product |
| **IDB** | Inter-American Development Bank |
| **LAC** | Latin America and the Caribbean |
| **NRW** | Non-Revenue Water |
| **O&M** | Operation and Maintenance |
| **PBCs** | Performance-Based Contracts |
| **SDGs** | Sustainable Development Goals |
| **UN** | United Nations |
| **WUE** | Water Use Efficiency |
| **WHO** | World Health Organization |
| **USD** | United States Dollar |
| **EPR** | Extended Producer Responsibility |
| **OECD** | Organisation for Economic Co-operation and Development |

# Executive summary

## Key findings

The circular water economy aims at reducing water use and enhancing water efficiency in production and consumption processes, reusing and recycling water, and recovering energy and materials from wastewater treatment. Globally, three main factors are driving the move towards circular water solutions: *megatrends affecting water quality, quantity and infrastructure resilience*, such as climate change, demographic growth and rapid urbanisation, heightening water demand and stressing existing infrastructure; *technological advancements and digitalisation*, allowing circular water business models to become more attractive and viable for water operators; and *socio-economic costs* of the predominant linear approach of "take, make and dispose", including impacts on the health and economic welfare of communities and on the viability of businesses dependent on water resources.

The Latin American region faces a complex array of challenges in relation to water resources and services. First, despite having the highest share of renewable water resources globally, the region is highly exposed to the impacts and costs of water-related disasters such as floods, droughts, and storms. Second, access to safe drinking water and sanitation remains a pressing concern. In 2020, one-quarter of the population in Latin America and the Caribbean lacked access to safely managed drinking water, meaning a reliable, onsite water source free from contamination. Additionally, two-thirds of the population, especially in rural areas, did not have access to safely managed sanitation, which requires private sanitation facilities that safely dispose of or treat waste. Third, 54% of domestic wastewater flow in the region is untreated, compared to 14% in Europe and North America and 42% globally. This lack of treatment leads to medium to extremely high physical water risks, with rural areas disproportionately affected: while 40% of household wastewater is treated in urban areas, only 9% is treated in rural communities.

This report builds on the OECD/Inter-American Development Bank (IDB) Survey on Water and Circular Economy to shed light on the circular water economy and identify the main challenges to the transition in ten Latin American countries, namely Argentina, Brazil, Chile, Colombia, Costa Rica, Honduras, Mexico, Paraguay, Peru, and Uruguay. The survey shows that 8 countries out of 10 consider water scarcity as the main driver to transition to the circular water economy. In addition, 6 out of 10 consider the circular water economy as an opportunity to address water pollution, ensure access to water and sanitation, and adapt to climate change.

The survey highlights that most countries predominantly adhere to a linear model for drinking water supply, sanitation and water resources management. Among the ten surveyed countries, seven have already implemented broad circular economy policies, while Mexico is currently developing one. Encouragingly, six countries – Brazil, Chile, Colombia, Costa Rica, Peru, and Uruguay – explicitly include water resources in their circular economy policies. The survey shows that when water is embedded in circular economy policies, measures mostly relate to reducing water use, reuse and recycling and, to a lesser extent, the recovery of materials and energy. For example, Brazil's Circular Economy Roadmap emphasises the need to optimise water use through water recycling technologies in both rural and urban settings. In Chile, the Circular Economy Roadmap includes specific measures for water resources, such as drafting a regulation for greywater reuse systems. Costa Rica includes water-related goals in its National Circular Economy Strategy and its Circular Economy Bill, stressing consumption reduction and wastewater treatment. Peru highlights efficient water resource use through initiatives such as the Circular Economy Pact and the National Circular Economy Roadmap.

Three main types of governance obstacles impede the transition towards circular water economies. First, inadequate regulatory frameworks and misalignments within and across policy areas emerge as prominent obstacles in Argentina, Brazil, Costa Rica, and Mexico. Institutional fragmentation and a lack of co-ordination among actors in water, waste and energy policy hinder the implementation of coherent circular economy initiatives in Paraguay and Peru. Second, insufficient financial resources hinder access to the technologies, infrastructure and capacities required for the reuse of by-products in Chile, Colombia, and Uruguay. Finally, the lack of awareness and education among key stakeholders, including policymakers, operators and civil society, the main barrier for Honduras, underscores the need for targeted outreach and capacity-building efforts to promote circular water management practices.

## Key recommendations

Building on the Survey results and the OECD Principles on Water Governance, this report presents a Roadmap to transition to the Circular Water Economy in Latin America composed of three key governance areas with 13 concrete actions:

- ***Strengthen regulatory and policy frameworks,*** by: (1) updating and streamlining regulations, particularly in the areas of material and energy recovery, and explicitly embedding circular economy principles into existing water policies to enhance urban wastewater management, water reuse, and sewage sludge management; (2) integrating circular water criteria into public procurement for by-products, and focusing on identifying and promoting markets for treated wastewater, biogas, and biosolids; (3) encouraging experimentation through regulatory innovation, such as sandbox regulation, allowing for temporary exceptions to current regulatory practices; and (4) adopting a systems approach that mainstreams circular water economy principles across sectors – including water and sanitation, agriculture, energy, urban and regional development, and industry – to ensure cohesive and effective policy alignment.
- ***Adapt financing and economic tools to promote innovative business models,*** through: (5) assessing and reducing investment needs, prioritising cost-effective measures such as reducing non-revenue water (NRW), which can be more energy-efficient and economical than building new infrastructure; (6) applying a life-cycle approach to investment decisions to reduce operational and maintenance costs while pinpointing externalities and generating revenues through the recovery of biosolids, nutrients, and energy; (7) diversifying financial sources, including engaging the private sector under transparent regulatory frameworks to ensure sustainable financing; (8) collaborating with regional development banks to close infrastructure gaps and promote innovation and research; (9) allocating financial resources to innovation in circular economy practices; and (10) incentivising regenerated materials and energy through economic mechanisms such as tax discounts, environmental taxes, or differentiated tariffs.
- ***Engage stakeholders and build capacity,*** through: (11) raising awareness and promoting effective stakeholder engagement in circular water practices through transparent communication and dissemination of accurate information, (12) ensuring that water-related decision-makers and practitioners are equipped with the skills needed to shift the paradigm; and (13) strengthening transparency and accountability of circular water systems and related data by quantifying key metrics such as NRW, water savings from conservation techniques, volumes of treated wastewater reused, recovered sludge repurposed as fertilisers, and energy savings achieved through recovery during treatment.

**Infographic 1. An overview of the OECD Report on the Circular Water Economy in Latin America**

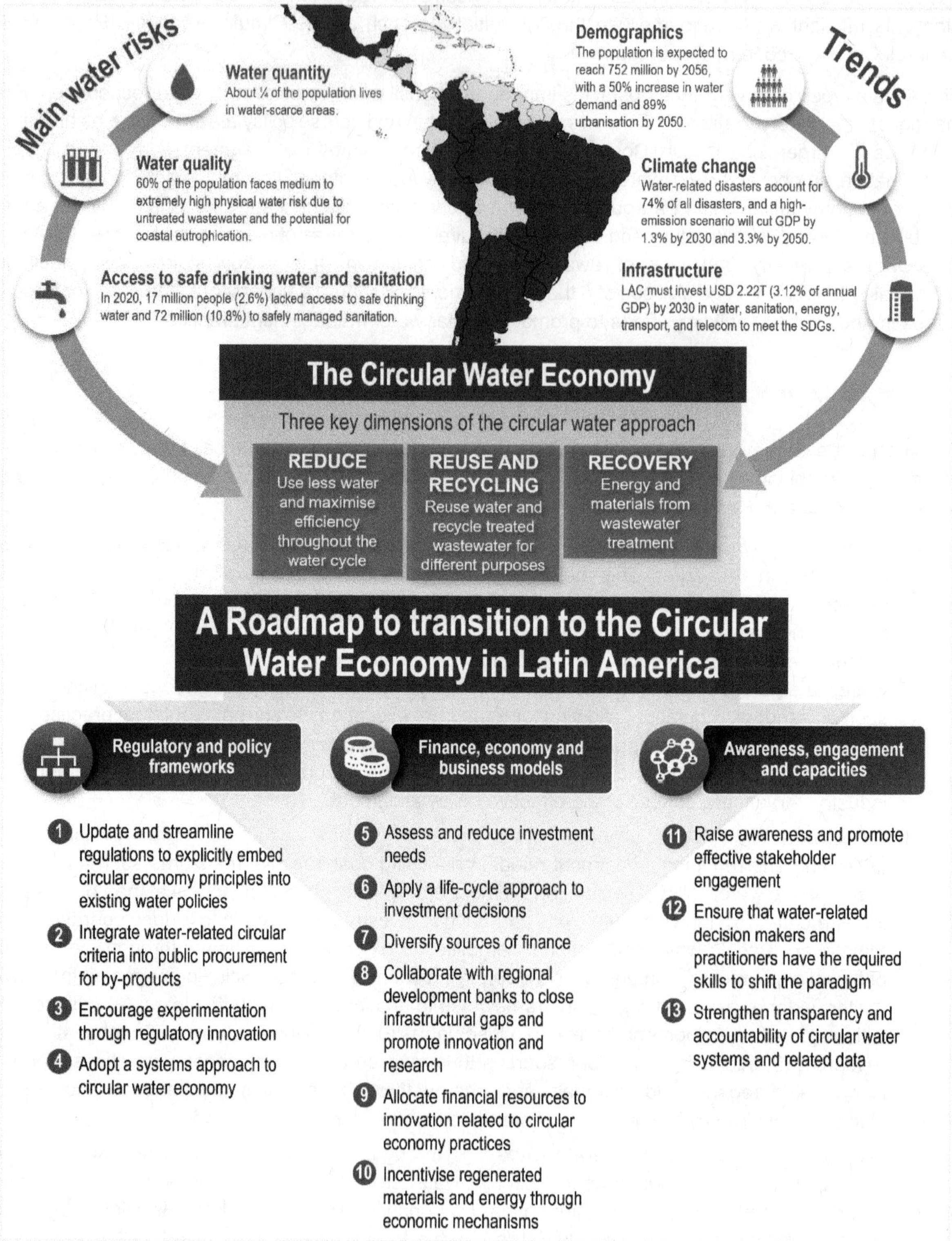

# 1 The Circular Water Economy in Latin America

This chapter establishes the framework for the circular water economy in Latin America by examining the region's key water challenges, identifying the primary drivers and opportunities for implementing such an approach and outlining the key dimensions of the circular water economy.

## Introduction

While there is no standard definition of the circular water economy[1], this report draws on the existing literature to define the circular water economy along three key dimensions: a) reducing water use and increasing water efficiency; b) increasing water reuse and recycling of treated wastewater; and c) recovering energy and materials from wastewater treatment. These dimensions align with the three key objectives of the broader concept of the circular economy (Ellen MacArthur Foundation, 2019[1]):

- Minimising waste and pollution by treating wastewater to produce clean water and recovering energy during treatment, which significantly reduces greenhouse gas emissions;
- Keeping resources in use in the economy for as long as possible, through water reuse and recycling, thereby reducing overall consumption and reliance on freshwater sources; and
- Regenerating nature by lowering freshwater extraction, conserving ecosystems, and ensuring water security (Figure 1.1).

The circular water economy provides economic benefits to both private companies and society by reducing operational costs through energy recovery and resource efficiency, minimising water costs by extending water reuse cycles, ensuring adequate water supplies, while managing the environmental impacts associated with water lifecycle and supply chains, and preserving natural resources to remain available for future generations (OECD, 2023[2]). Addressing water scarcity, service disruptions and pollution requires bold, transformative action from both public and private sectors, including in Latin America, a region that faces a unique set of water challenges. Climate change, coupled with rapid urbanisation and demographic growth, is putting increasing pressure on existing water infrastructure and services. Without urgent and coordinated action, the risks of water scarcity and quality degradation will continue to grow, compromising both the health of communities and the viability of businesses dependent on water resources.

After providing a detailed overview of water challenges and risks in Latin America, this chapter will detail each of the aforementioned dimensions of the circular water economy, with relevant examples in Latin America and in the world.

**Figure 1.1. The Circular Water Economy framework**

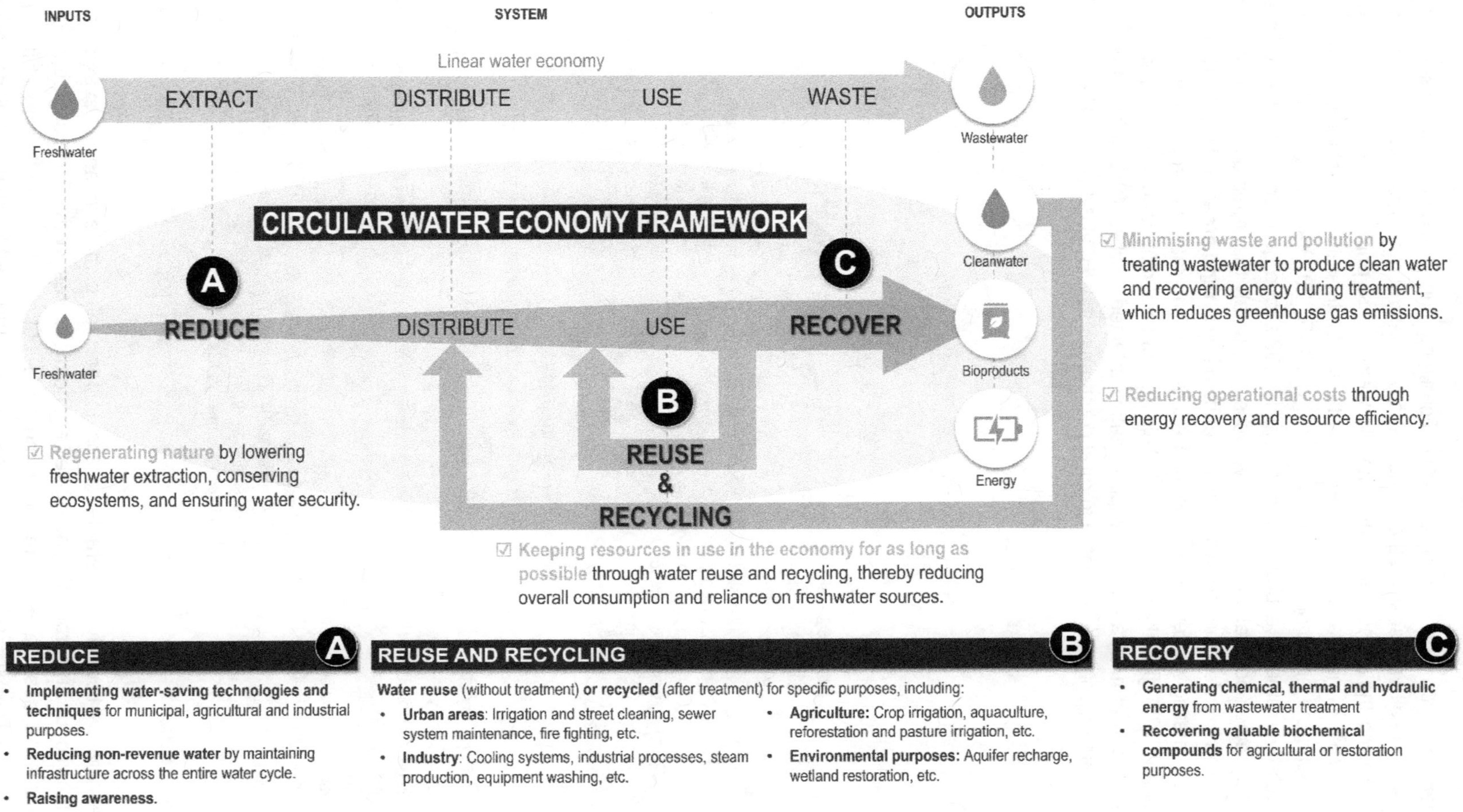


Source: Author's elaboration

## Why it matters: water challenges in Latin America

The Latin America[2] region faces a complex array of challenges in relation to water resources and services (Figure 1.2). First, despite possessing the highest share of natural renewable water resources globally, the region stands among the most vulnerable ones to water risks, primarily due to the impacts of climate change. Around 75% of natural disasters are climate and water-related, including floods, droughts, and storms, which generate enormous economic costs (UNDRR, 2023[3]). Second, access to safe drinking water and sanitation remains a pressing concern. In 2020, one-quarter of the Latin America and the Caribbean (LAC) region population did not have access to safely managed drinking water, and two-thirds did not have access to safely managed sanitation, particularly in rural areas (WHO/UNICEF, 2022[4]). Third, in 2022, 54% of domestic wastewater flow in the LAC region is untreated, compared to 14% in Europe and North America and 42% globally (UN, 2024[5]). This lack of treatment leads to medium or extremely high physical water risk, with rural areas disproportionately affected – while 40% of household wastewater is treated in urban areas, only 9% is treated in rural communities (WHO/UNICEF, 2022[4]).

### Figure 1.2. Main water risks and trends in LAC

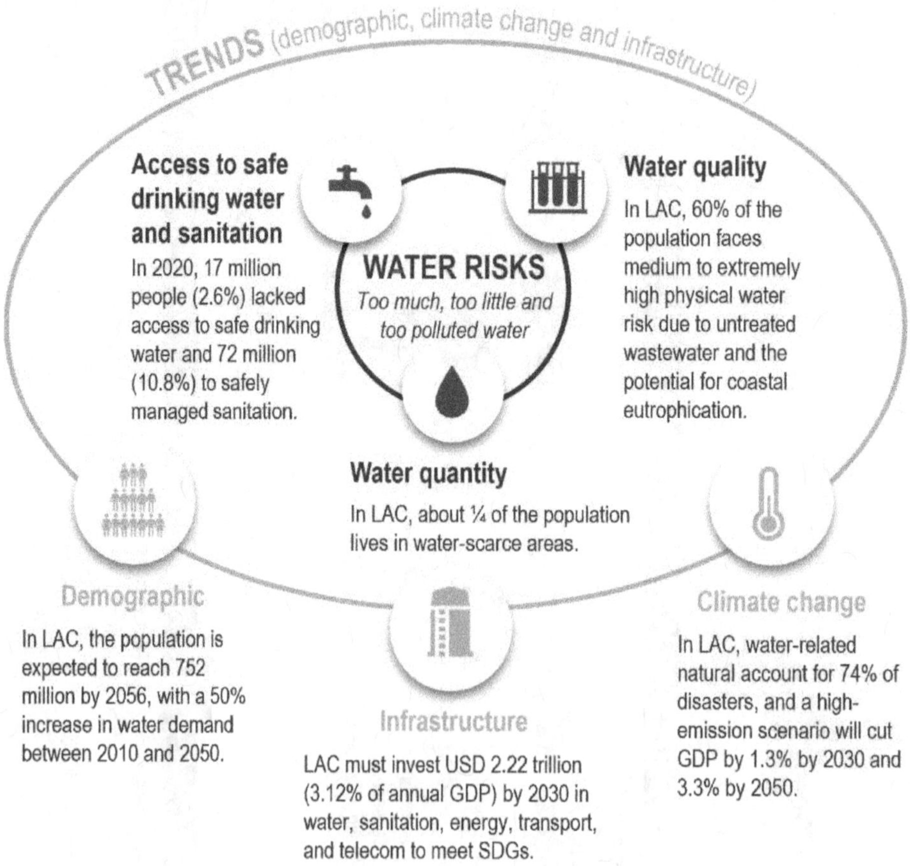


Source: Author's elaboration

LAC countries are particularly vulnerable to the impacts of climate change, experiencing recurrent disasters such as floods and droughts and related economic losses. With 35.1% of global renewable water resources, LAC is the most water-rich region in the world, but also amongst the most affected regions by climate change (OECD et al., 2022[6]). A high-emission scenario would lead to a regional reduction in GDP per capita of 1.3% by 2030 and 3.3% by 2050, compared to a scenario where temperatures do not rise

(Van Der Borght et al., 2023[7]). A total of 74% of natural disasters in the LAC region are water-related, including floods, droughts, and storms. Between 2000 and 2022, floods ranked as the most prevalent disaster in the region (44% of the natural disasters), affecting Brazil, Colombia and Peru in particular. Floods resulted in over USD 1 billion in total damages on 12 occurrences during this period. Droughts have impacted over 53 million people in the region (UNDRR, 2023[3]; EM-DAT, (n.d.)[8]), and incurred USD 13 billion in losses due to declines in agricultural production between 2008 and 2018 (FAO, n.d.[9]). One-quarter of the population (150 million people) live in water-scarce areas (World Bank, 2022[2]). Freshwater availability per person in Central America is predicted to plummet by at least 82% by 2100 compared to 2005 levels, with a projected 11% decrease in total rainfall by 2050 (ECLAC, 2010[10]), raising serious concerns about water and food security (IPCC, 2022[11]).

Universal coverage of safe drinking water and sanitation remains a pressing concern in the LAC region. Between 2000 and 2020, approximately 164 million additional individuals had gained access to basic drinking water services (with 144 million in urban areas and 20 million in rural areas), and approximately 195 million to basic sanitation services (with 167 million in urban areas and 29 million in rural areas) (WHO/UNICEF, 2022[12]). Despite progress, 17 million people (2.6% of the LAC population) still lack access to safely managed drinking water services, and 72 million people (10.8%) remain without access to safely managed sanitation services in the region (WHO/UNICEF, 2022[12]). In urban settings, 19% of the population lack access to safely managed drinking water, and 40% to safely managed sanitation services, while in rural areas, nearly half of the population (47%) lack access to safely managed drinking water. In rural areas of countries such as Peru, up to 15% of the population still lack access to basic drinking water services, and more than one-third of the population still lack access to basic sanitation services (Figure 1.3 and Figure 1.4).

## Figure 1.3. Drinking water services coverage in urban and rural areas in surveyed Latin American countries

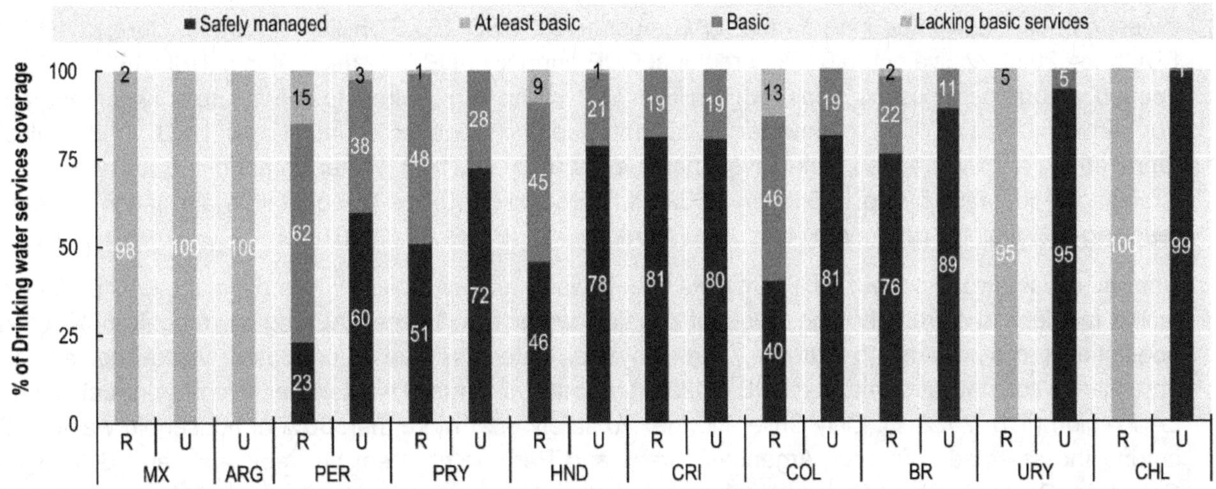


Note: Data from 2022 distinguishes urban (U) and rural (R) areas in access to water. Basic drinking water services refer to water from improved sources, accessible within 30 minutes round trip. Improved sources include piped water, boreholes, protected wells/springs, and packaged or delivered water. "At least basic" includes both basic and safely managed water services.
Source: Author's elaboration based on data from WHO/UNICEF (2024) Joint Monitoring Programme (JMP) for Water Supply, Sanitation and Hygiene, https://washdata.org/data/household#!/

Figure 1.4. Sanitation services coverage in urban and rural areas in surveyed Latin American countries

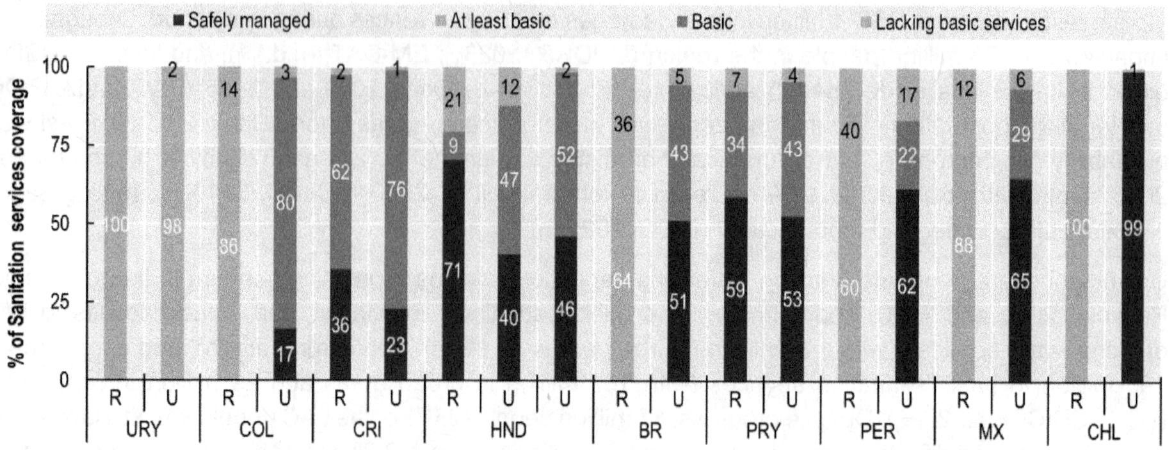


Note: Data from 2022 distinguishes urban (U) and rural (R) areas in access to water. "At least basic" indicates the percentage of people using improved sanitation facilities that are not shared with other households. Improved sanitation facilities include flush/pour flush to piped sewer systems, septic tanks or pit latrines; ventilated improved pit latrines, compositing toilets or pit latrines with slabs.
Source: Author's elaboration based on data from WHO/UNICEF (2024) Joint Monitoring Programme (JMP) for Water Supply, Sanitation and Hygiene, https://washdata.org/data/household#!/

Existing water infrastructure falls short of ensuring universal drinking water and sanitation coverage, while population growth and urbanisation are exacerbating pressure on water demand and supply. A total of 7.2% of total investment (public and private) from 1990 to 2018 in LAC was allocated to essential services, including electricity, gas, and water (ECLAC, 2022[13]; OECD et al., 2023[14]). In 2021, investment in water infrastructure in LAC accounted for 0.26% of the region's GDP (Infralatam, 2021[15]), which is among the lowest globally. Estimates indicate that LAC must invest USD 255.97 billion in infrastructure to achieve SDG6[3] by 2030, equivalent to 0.5% of regional GDP annually (IDB, 2021[16]). Of this, USD 183.8 billion is needed to address existing gaps, while USD 72.2 billion is required to meet future water demand. Furthermore, closing the wastewater treatment gap will require an additional USD 16.85 billion, representing a 7% increase in total investment needs (IDB, 2021[16]). Water demand in South America is expected to increase by up to 50% by 2050 compared to 2010 due to population and income growth, reaching 5% of global demand by that date (currently 4%) (Burek et al., 2016[17]).

Untreated wastewater poses significant risks to both water and air quality, with most of the LAC population facing medium to extremely high levels of physical water risk. In the LAC region, roughly 60% of the population faces medium to extremely high physical water risk due to untreated wastewater and the potential for coastal eutrophication (IDB, 2022[18]). About two-thirds of wastewater remains untreated in the LAC region (IDB, 2022[18]). Only Chile and Mexico safely treat more than 60% of household wastewater among the selected countries. Argentina, Brazil and Peru safely treat between 35% and 60%, while Colombia, Costa Rica and Paraguay manage between 15% and 30%. Honduras and Uruguay currently lack monitoring systems for this indicator (Figure 1.5). Wastewater systems are also responsible for generating greenhouse gas (GHG) emissions, both directly through the breakdown of excreta released into the environment or during treatment processes, and indirectly through the energy needed for treatment steps Box 1.1).

**Figure 1.5. Percentage of domestic wastewater flows safely treated in surveyed Latin American countries**

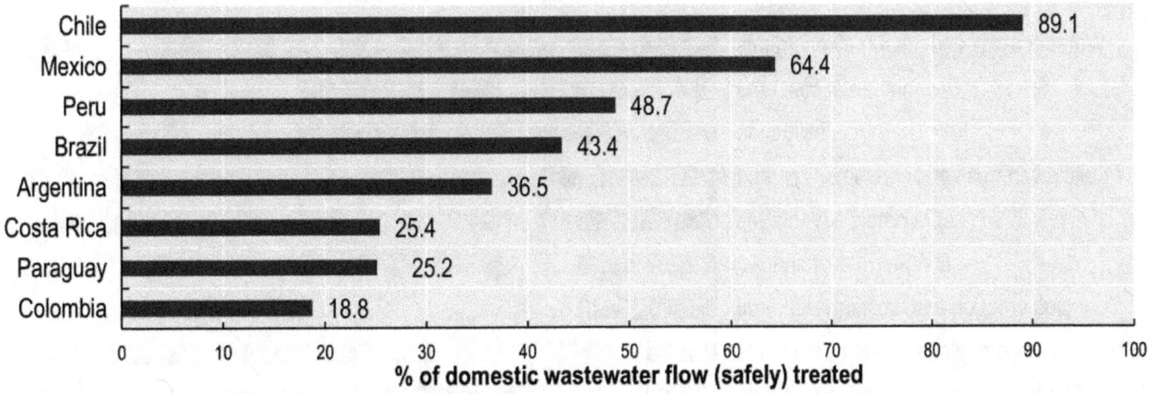


Note: Data from 2022. No data is available for Honduras and Uruguay.
Source: Authors' elaboration based on data from UN (2024), Progress on Wastewater Treatment (SDG target 6.3), https://www.sdg6data.org/en/indicator/6.3.1

### Box 1.1. Greenhouse gas emissions from the water and sanitation sector

Water use, storage and distribution are responsible for approximately 10% of global greenhouse gas (GHG) emissions. The degradation of organic material during wastewater treatment contributes to 1.57% of global GHG emissions and 5% of global non-carbon dioxide GHG emissions, primarily methane and nitrous oxide. Moreover, treatment plants are estimated to consume about 3% of global electricity. These emissions are expected to increase. By moving towards SDG targets 6.2 (Sanitation and hygiene) and 6.3 (Progress on Wastewater Treatment), there is potential to lower GHG emissions from sanitation and wastewater systems by harnessing energy and nutrients found in wastewater. Applying a circular approach to the water sector can reduce GHG emissions by generating renewable energy during the wastewater treatment process and replacing fossil-based fertilisers. This process decreases environmental pollution while bolstering food production through nutrient recovery and enhancing energy security by reducing reliance on conventional power sources.

Source: CDP (2020), A wave of change: The role of companies in building a water-secure world, https://www.cdp.net/en/research/global-reports/global-water-report-2020; Lu, L. et al. (2018), Wastewater treatment for carbon capture and utilization", Nature Sustainability, https://doi.org/10.1038/s41893-018-0187-9

According to the OECD/IDB Survey on Water and Circular Economy (Box 1.2), water scarcity and resource constraints are key drivers in moving from a linear to the circular water economy by 80% of the ten surveyed countries, followed by environmental concerns and risk mitigation (60%) (Figure 1.6). Argentina, Brazil, Honduras, Mexico, and Uruguay consider water scarcity and resource constraints as the top drivers, while environmental concerns rank first for Colombia, Costa Rica, and Paraguay, and risk mitigation and regulatory pressures are the respectively the most prominent for Chile and Peru. Addressing water pollution, ensuring access to water and sanitation, and adapting to climate change are the primary opportunities identified by 60% of surveyed countries (Figure 1.7).

Figure 1.6. Drivers for the circular water economy in surveyed Latin American countries

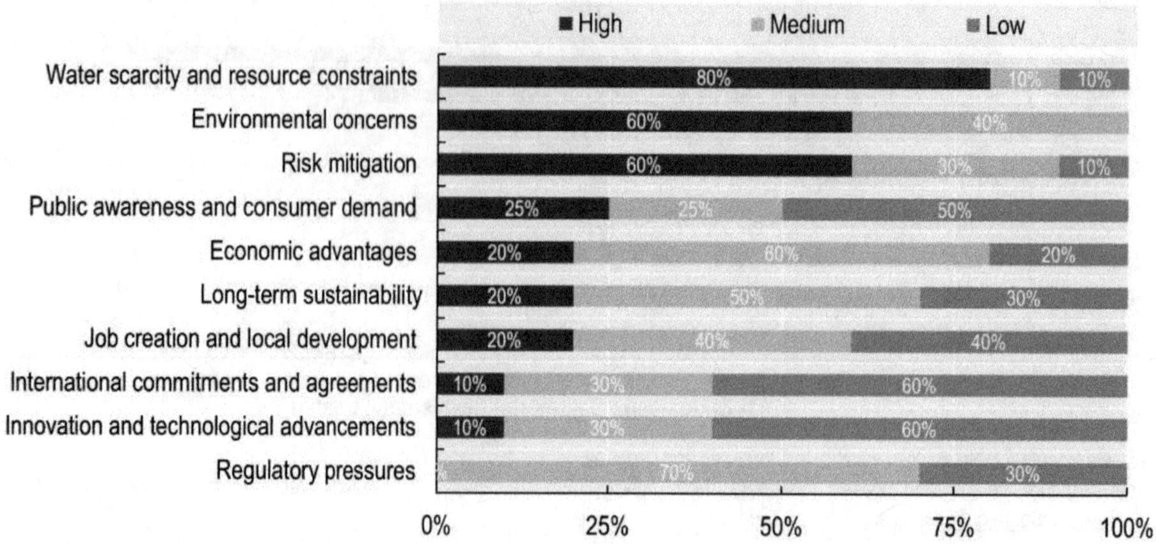


Note: Countries ranked each of the drivers on a scale of 1-10, with 10 being the highest score. The high score category corresponds to a score between 8-10, followed by the medium score category (4-7) and the low score category (1-3).
Source: OECD/IDB (2024) OECD/IDB Survey on Water and Circular Economy in Selected Countries of Latin America.

Figure 1.7. Opportunities for the circular water economy in surveyed Latin American countries

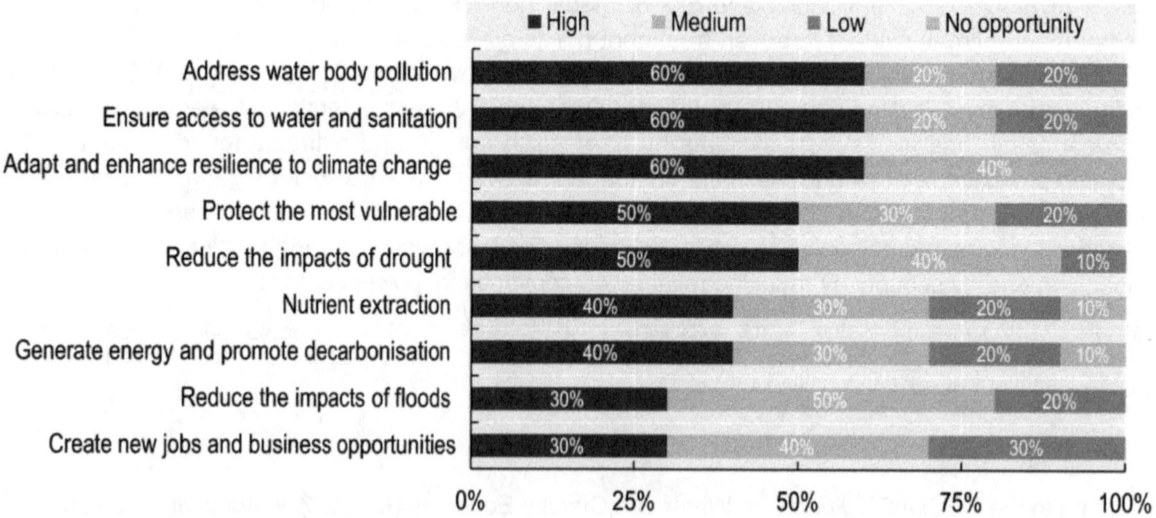


Note: Countries classified opportunities as 'Great opportunity', 'Medium opportunity', 'Low opportunity', and 'Absence of opportunity'.
Source: OECD/IDB (2024) OECD/IDB Survey on Water and Circular Economy in Selected Countries of Latin America

**Box 1.2. OECD/IDB Survey on Water and Circular Economy**

The OECD/IDB Survey on Water and Circular Economy, carried out between October 2023 and February 2024, targeted ten countries in Latin America: Argentina, Brazil, Chile, Colombia, Costa Rica, Honduras, Mexico, Paraguay, Peru, and Uruguay (Figure 1.8). The survey was structured around five sections:

1. Identify the key drivers of the water circular economy.
2. Map the roles and responsibilities related to the circular water economy across levels of government.
3. Identify the main water and sanitation and circular economy legislations, regulations and policies and how they systematically address the circular water economy focusing on specific dimensions such as water use reduction, reuse, recycling, and the recovery of energy and materials.
4. Identify initiatives in the water sector related to the circular economy and clarify aspects such as stakeholder engagement, financing, and data and assessment measures.
5. Identify the challenges, opportunities and priority governance dimensions needed to transition from a linear to the circular water economy.

**Figure 1.8. Geographical coverage of the Survey**

Source: Author's elaboration.

## The circular water economy in detail

The concept of the circular water economy focuses on three key dimensions: a) reducing water use and increasing water efficiency, b) promoting water reuse and recycling of treated wastewater, and c) recovering energy and materials from wastewater treatment. These concepts align with the broader circular economy goals of minimising waste and pollution, keeping resources in use for as long as possible, and regenerating nature. These dimensions are presented in more detail below.

Reducing water use is the first step in the circular water economy. This can be achieved by implementing advanced water-saving technologies across municipal, agricultural, and industrial sectors to increase water use efficiency, reducing non-revenue water (NRW) through the maintenance and upgrading of infrastructure, and raising awareness about the importance of water conservation. Rainwater harvesting systems, for example, can capture and store rainwater for non-potable uses like landscaping, cleaning, or toilet flushing, reducing dependence on freshwater supplies. As a result, water use can be significantly curtailed at the source, laying the foundation for a sustainable cycle.

When water is used, efforts should shift toward reuse. In some cases, water can be reused without treatment, making the most of its residual quality. In urban settings, greywater from sinks or showers can be reused directly for irrigating gardens or toilets flushing. In agriculture, water used for cleaning fruits and vegetables can be redirected for crop irrigation or livestock watering. Industrial processes can use cooling water for washing equipment or dust suppression.

Where direct reuse is not feasible, wastewater should be treated and recycled. When treated to the appropriate standards, wastewater's quality can be restored to a level that allows it to be reinjected into the distribution system. This effectively reduces or, in some cases, replaces the need for freshwater extraction. For reuse to a potable water standard, advanced treatment processes like microfiltration, reverse osmosis, and UV disinfection with advanced oxidation can purify wastewater to meet drinking water standards, enabling it to be directly supplied to consumers or reintroduced into natural reservoirs or aquifers for later use. For non-potable purposes, treated wastewater can be used for agricultural irrigation, industrial cooling, or landscaping, significantly conserving freshwater resources. Additionally, treated wastewater can replenish wetlands, rivers and groundwater systems.

As a result of the wastewater treatment process, sludge is obtained in addition to treated water. Residual sludge contains high value materials including phosphorus, nitrogen, and sulphur that can be recycled, reducing the demand for virgin resources (Solon, 2019[19]). Biosolids (sludge treated to levels that permit its beneficial use) can be used to recover degraded land and as compost in agriculture (IEA, 2015[20]), helping mitigate water pollution, while decreasing costs for farmers (World Bank, 2019[21]). Using biosolids instead of disposing of them in landfills, also lowers or eliminates transport and landfill costs for water utilities, while reducing GHG emissions (UN World Water Assessment Programme, 2017[22]; Waternet, 2017[23]; World Economic Forum, 2020[24]). The remaining biomass can be incinerated to produce sewage sludge ash which, when used as agricultural fertiliser, produces similar comparable yield results to conventional phosphate fertiliser (Franz, 2008[25]). Sewage sludge ash can be employed in the construction industry as a replacement aggregate for use in concrete and mortar to make bricks or tiles or incorporated as raw material for cement (Smol, 2015[26]). Pilot projects are underway to recover cellulose from wastewater (EurEau, 2021[27]).

In addition, wastewater contains energy that can be captured in a number of forms, including thermal, chemical and hydraulic energy. Thermal energy is about capturing heat, used for district heating and cooling, powering agricultural greenhouses, and drying sludge. Chemical energy is stored in organic compounds present in wastewater and it involves converting organic compounds into usable fuels. Hydraulic energy focuses on harnessing the movement of freshwater or wastewater for power generation.

Each type of energy has specific methods of recovery and unique applications. Estimates of the recoverable energy embedded in municipal wastewater suggests that the potential for thermal energy (80% of energy recovered) is much higher than for chemical energy (20%). Only a very small amount (less than 1%) of the embedded energy is in the form of hydraulic energy (Tarallo, 2014[28]). Studies have demonstrated that wastewater contains five to ten times more energy than the energy needed for treatment. While only some of this energy can be recovered, it is possible for the largest urban wastewater treatment plant to be net energy producers (Riley and alii, 2020[29]). In current practices, the energy potential of wastewater is not fully exploited and, although several energy-neutral or energy-positive plants exist and operate fully, they are not yet the norm. Wastewater treatment plants can become 100% self-sufficient in energy terms if they effectively employ energy efficiency and energy harvesting from wastewater.

## The potential for circular water economy in Latin America

According to the OECD/IDB Survey, most Latin American countries predominantly adhere to a linear model for water supply, sanitation and resources management. Only Colombia, Mexico, and Peru perceive their water system as partially circular. As a result, there is considerable potential for countries in the Latin America to dramatically increase the circularity of their water systems. This section details the identified dimensions of the circular water economy and illustrates the potential for Latin American countries, drawing in the experiences of other countries and regions as well as the results of the OECD/IDB Survey.

### *Reducing water use and using it efficiently*

Water can be lost in networks and used inefficiently. Globally, leaks and physical pipe breakages lead to the loss of over 33 billion m$^3$ of treated water every year. In developing countries, roughly 45 million cubic metres of water are lost daily with an economic value of over USD 3 billion per year (World Bank, 2016[30]). Saving half of those losses would provide enough water to serve at least 90 million people. According to OECD (2016[31]), smaller cities surveyed (under 1.5 million inhabitants) reported higher average water loss than larger cities. The correlation between GDP per capita and the share of water losses shows greater water losses in cities with lower GDP per capita. Wastage in these cities is generally associated with unauthorised consumption, poor connections and metering inaccuracies (Farley, 2001[32]). A substantial number of cities across Latin America continue to face high levels of physical water losses. In the LAC region, the NRW rate exceeds 40% (IDB, 2018[33]).

The economically optimal level of water losses in municipal networks is estimated between 10% and 20% on average, depending on the nature of individual systems (OECD, 2016[31]). According to an approach adopted by the European Commission, this level is reached at the point at which the cost of reducing leakage is equal to the benefit gained from further leakage reductions (European Commission, 2013[34]). Appropriate leakage management can help reduce significantly water abstraction, recover revenue from water losses and in some cases mitigate the need for water source expansion (US Environmental Protection Agency, 2016[35]). Reducing water losses can generate economic value for utilities but requires strong investments. For example, in the European Union (EU), the overall cost of reduce water losses by 10% amounts to approximately EUR 8 billion, and about twice the amount would be needed to achieve a 20% leakage rate, underlining the financial issues at stake to reach low levels of leakage (OECD, 2023[36]).

Water-saving techniques and awareness raising are important to increase water use efficiency. In 2021, global water use efficiency (WUE)[4] stood at USD/m$^3$ 20.8, and data indicates a positive trajectory, with water-use efficiency increasing from USD/m$^3$ 17.3 in 2015 to 18.9 USD/m$^3$ in 2018, reflecting a 20% efficiency gain (FAO/UN Water, 2021[37]). Several countries are establishing targets and providing solutions. For example, Germany's 2023 National Water Strategy emphasises water-saving technologies, better wastewater management, and sustainable farming practices. In 2023, France unveiled a comprehensive Water Saving Plan, targeting a 10% reduction in national water use by 2030. Among surveyed Latin American countries, in 2021, only Brazil surpassed global WUE, with Costa Rica closely following. However, three out of the ten countries – Chile, Colombia and Peru– fall short of achieving even half of the global water efficiency level (Figure 1.9).

Figure 1.9. Water use efficiency in surveyed Latin American countries

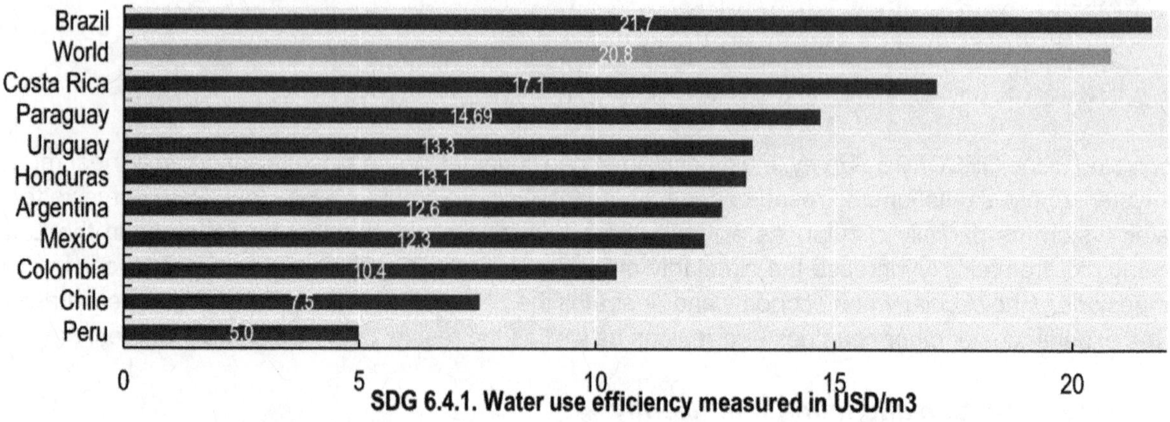


Note: Data from 2021. Water Use Efficiency (WUE) is the sum of the efficiencies in the major economic sectors weighted according to the proportion of water withdrawn by each sector over the total withdrawals.
Source: Author's elaboration based on FAO AQUASTAT database https://data.apps.fao.org/aquastat/?lang=en&share=f-2a4f649d-bc55-4fddbeea-734145de5a00.

### *Reusing water and recycling treated wastewater*

Water reuse refers to the direct repurposing of used water without undergoing treatment. This practice is often limited to non-sensitive applications where the quality of the water is already adequate for the intended use. Globally, only 11% of wastewater is reused, highlighting significant room for increased reuse (UNEP, 2023[38]). In agriculture, particularly in water-scarce regions like the Middle East and North Africa, reuse plays an important role in supporting irrigation needs. In LAC, the potential for reuse is underexploited despite agriculture being the primary consumer of water. In the region, where irrigation accounts for over 60% of water withdrawals on average (Figure 1.10), reuse rates for untreated water remain negligible. For instance, in surveyed Latin American countries, reuse of untreated wastewater is almost non-existent except in isolated cases where minimal-quality water can meet non-drinking needs. Reuse typically involves informal practices without systemic implementation.

Figure 1.10. Water withdrawals by sector in surveyed Latin American countries

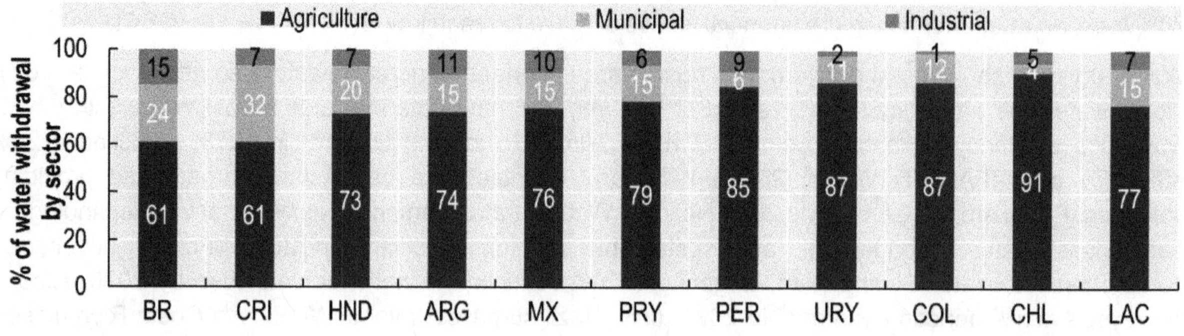


Note: The data correspond to 2021
Source: Author's elaboration based on FAO AQUASTAT database https://data.apps.fao.org/aquastat/?lang=en&share=f-2a4f649d-bc55-4fddbeea-734145de5a00.

Water recycling, on the other hand, refers to the process of treating used water at various levels (primary, secondary, or advanced) to a quality that is suitable for industrial processes, irrigation, or even potable water supply. Globally, around 359 billion m³ of wastewater is produced annually, yet 48% is discharged untreated (Jones et al., 2020[39]). Increasing recycling rates could address water scarcity and environmental concerns.

In Latin American countries, the potential of wastewater recycling is still to be unlocked. Chile and Mexico safely treat over 60% of household wastewater, but recycling rates remain low, with Mexico leading at 20%. Argentina, Brazil and Peru treat 35% to 60% of wastewater, yet recycling efforts focus more on industrial rather than municipal uses (UN, 2024[40]). Opportunities abound for increasing recycling rates by expanding wastewater treatment infrastructure. Treated wastewater could significantly supplement agriculture and industrial water needs, addressing shortages. Some examples are already in place: the Atotonilco de Tula wastewater treatment plant in Mexico recycles water for irrigation purposes, while the Aquapolo wastewater treatment plant in São Paulo, Brazil, recycles water specifically for industrial use.

In Europe, water reuse and recycling are structured practices with specific applications. Approximately 1 billion out of 40 billion m³ of treated wastewater is reused annually, though the European Commission estimates this figure could be increased sixfold (Water Reuse Europe, 2020[41]). Out of the almost 800 water reuse and recycling practices identified in Europe, 62% are concentrated in water-scarce countries like Spain, while coastline regions with high tourism pressure prioritise treated water reuse and recycling to counteract resource depletion (Water Reuse Europe, 2020[41]). Overall, agricultural reuse remains the most common water reuse application in Europe (39%) followed by industrial reuse (15%) and reuse for recreational purposes (11%) (Water Reuse Europe, 2020[41]).

Outside Europe, the case of Singapore is often mentioned as a best practice. In 2003, the Public Utilities Board, Singapore's National Water Agency, introduced NEWater, a high-grade reclaimed water produced from treated used water, which exceeds the drinking water standards set by the World Health Organization (WHO) and the US Environmental Protection Agency. NEWater is used primarily for non-potable industrial purposes at wafer fabrication parks, industrial estates and commercial buildings (OECD, 2016[31]). After almost 20 years in action, NEWater supplied 30% of Singapore's demand for water in 2022, with the aim of covering 50% by 2060 (Rahmawan and Eliana, 2023[42]).

### *Recovering energy and materials from wastewater treatment*

As noted above, recovering energy and materials from wastewater treatment is a core part of the circular water economy, but is still in its infancy in most parts of the world. Advanced technologies allow for the recovery of biogas (a source of energy) and valuable biochemical compounds.

Bio-factories are particularly innovative and have transformed the concept of a wastewater treatment plant by introducing the recovery of materials and energy. In the metropolitan region of Santiago, Chile, for example, Aguas Andinas set three bio-factories: La Farfana, Mapocho-Trebal, and La Florida. These bio-factories collectively treat 100% of Greater Santiago's wastewater, allowing a significant portion of clean water to be reintroduced into the Mapocho River, with the remainder allocated for irrigation in the metropolitan region. They won the United Nations "Momentum for Change Climate Action Award" in 2018. Beyond wastewater treatment, the goal of these bio-factories is to generate zero waste, be self-sufficient in energy and carbon neutral by extracting and providing resources such as electricity, natural gas, agricultural fertiliser, and clean water (IDB, 2022[43]).

In another example, the bio-factory in the city of Granada, Spain, aims at shifting from being a significant energy consumer to becoming energy-to-energy producers; recycling treated water rather than only purifying and returning it to the natural environment; and transforming waste into resources rather than sending it to landfill (OECD, 2021[44]). In 2019, the bio-factory almost reached its 100% energy self-sufficiency goal. Furthermore, 18.91 million $m^3$ of treated water have been reused for irrigation and for the maintenance of the minimum ecological flow of the local Genil River. In addition, from the 16 525 tonnes of fresh sludge material produced in the bio-factory in 2019, 14.3% was recycled for compost and 85.7% for direct application in the agricultural sector (OECD, 2021[44]).

## References


Burek, P. et al. (2016), *Water futures and solution-fast track initiative.*, https://pure.iiasa.ac.at/id/eprint/13008/. [17]

ECLAC (2022), *Economic Survey of Latin America and the Caribbean*, https://repositorio.cepal.org/server/api/core/bitstreams/e77f282a-c7cb-42bc-9045-d58e955e6504/content. [13]

ECLAC (2010), *The Economics of Climate Change in Central America: Summary 2010*, https://digitallibrary.un.org/record/717425?v=pdf (accessed on April 2024). [10]

Ellen MacArthur Foundation (2019), *Introduction to the Circular Economy*, https://www.ellenmacarthurfoundation.org/the-circular-economy-in-detail-deep-dive. [1]

EM-DAT ((n.d.)), *The international disaster database*, https://www.emdat.be/. [8]

EurEau (2021), *Products from Waste Water - Fibres, polymers and other organics*, https://www.eureau.org/documents/diverse/6080-factsheet-on-products-from-waste-water-fibres-and-polymers/file. [27]

European Commission (2013), *Resource and economic efficiency of water distribution networks in the EU*, http://ec.europa.eu/environment/water/quantity/pdf/Final%20REE%20Report%20Oct%202013.pdf. [34]

FAO (n.d.), *Damage and loss*, https://www.fao.org/resources/digital-reports/disasters-in-agriculture/en/ (accessed on April 2024). [9]

FAO/UN Water (2021), *Progress on change on water-use efficiency*, https://doi.org/10.4060/cb6413en. [37]

Farley, M. (2001), *Leakage Management and Control: A Best Practice Training Manual,*, http://whqlibdoc.who.int/hq/2001/WHO_SDE_WSH_01.1_eng.pdf. [32]

Franz (2008), "Phosphate fertilizer from sewage sludge ash", *Waste Management, Volume 28, Issue 10*, pp. 1809-1818, https://doi.org/10.1016/j.wasman.2007.08.011. [25]

IDB (2022), *El reúso de agua residual tratada en América Latina y el Caribe: 10 estudios de caso*, https://doi.org/10.18235/0004515. [43]

IDB (2022), *Open configuration options Scarcity in the Land of Plenty*, Inter-American Development Bank, https://doi.org/10.18235/0003969. [18]

IDB (2021), *La brecha de infraestructura en América Latina y el Caribe: estimación de las necesidades de inversión hasta 2030 para progresar hacia el cumplimiento de los Objetivos de Desarrollo Sostenible*, Inter-American Development Bank, https://doi.org/10.18235/0003759. [16]

IDB (2018), *How Efficient is the Supply of Water in Latin America?*, https://www.caf.com/en/currently/news/2018/03/how-efficient-is-the-supply-of-water-in-latin-america/#:~:text=CAF%20%2D%20development%20bank%20of%20Latin,in%20the%20supply%20of%20water. [33]

IEA (2015), *Annual Report*. [20]

Infralatam (2021), *Public Investment in Economic Infrastructure: Evolution by Country and Sector*, https://www.infralatam.info/home/. [15]

IPCC (2022), *Climate Change 2022: Impacts, Adaptation and Vulnerability, Intergovernmental Panel on Climate Change*, https://report.ipcc.ch/ar6wg2/pdf/IPCC_AR6_WGII_FinalDraft_FullReport.pdf (accessed on April 2024). [11]

Jones, E. et al. (2020), "Country-level and gridded wastewater production, collection, treatment and re-use [dataset]", *PANGAEA*, https://doi.org/10.1594/PANGAEA.918731. [39]

OECD (2023), "Circular economy - waste and materials", in *Environment at a Glance Indicators*, OECD Publishing, Paris, https://doi.org/10.1787/f5670a8d-en. [2]

OECD (2023), *Implementing Water Economics in the EU Water Framework Directive*, OECD Studies on Water, OECD Publishing, Paris, https://doi.org/10.1787/d6abda81-en. [36]

OECD (2021), *The Circular Economy in Granada, Spain*, OECD Urban Studies, OECD Publishing, Paris, https://doi.org/10.1787/5f8bd827-en. [44]

OECD (2016), *Water Governance in Cities*, OECD Studies on Water, OECD Publishing, Paris, https://doi.org/10.1787/9789264251090-en. [31]

OECD et al. (2023), *Latin American Economic Outlook 2023: Investing in Sustainable Development*, OECD Publishing, Paris, https://doi.org/10.1787/8c93ff6e-en. [14]

OECD et al. (2022), *Latin American Economic Outlook 2022: Towards a Green and Just Transition*, OECD Publishing, Paris, https://doi.org/10.1787/3d5554fc-en. [6]

Rahmawan, A. and G. Eliana (2023), "Mitigating Aquifer Crisis in Indonesia's New Capital, Nusantara: Problems and Lessons Learned from Singapore", *Közigazgatási és Infokommunikációs Jogi PhD Tanulmányok*, Vol. 4/1, pp. 5-22, https://doi.org/10.47272/KIKPhD.2023.1.1. [42]

Riley and alii (2020), "Techno-economic assessment of CHP systems in wastewater treatment plants", *Environments*, https://doi.org/10.3390/environments7100074. [29]

Smol, M. (2015), "The possible use of sewage sludge ash (SSA) in the construction industry as a way towards a circular economy", *Journal of Cleaner Production, 95*, pp. 45-54, https://doi.org/10.3390/en16093911. [26]

Solon, K. (2019), "Resource recovery and wastewater treatment modelling", *Environmental Science: Water Research & Technology, 5(4)*, pp. 631-642, https://doi.org/10.1039/C8EW00765A. [19]

Tarallo, S. (2014), *Utilities of the Future Energy Findings*, IWA Publishing. [28]

UN (2024), *Progress on Wastewater Treatment (SDG target 6.3)*, https://www.sdg6data.org/en/indicator/6.3.1. [5]

UN (2024), *Progress on Wastewater Treatment (SDG target 6.3)*, https://www.sdg6data.org/en/indicator/6.3.1. [40]

UN World Water Assessment Programme (2017), *The United Nations World Water Development Report 2017. Wastewater: The Untapped Resource*. [22]

UNDRR (2023), *Overview of Disasters in Latin America and the Caribbean 2000-2022*, https://www.undrr.org/media/89900/download?startDownload=20240503. [3]

UNEP (2023), *Wastewater – Turning Problem to Solution*, https://wedocs.unep.org/20.500.11822/43142. [38]

US Environmental Protection Agency (2016), *Best Practices to Consider when Evaluating Water Conservation and Efficiency as an Alternative for Water Supply Expansion*, https://www.epa.gov/sustainable-water-infrastructure/best-practices-water-conservation-and-efficiency-alternative-water. [35]

Van Der Borght, R. et al. (2023), "Los efectos del cambio climático en la actividad económica de América Latina y el Caribe", *CEPAL*. [7]

Water Reuse Europe (2020), *The state of the sector*, https://www.water-reuse-europe.org/about-water-reuse/#page-content. [41]

Waternet (2017), *Annual Report*. [23]

WHO/UNICEF (2022), *PROGRESS ON DRINKING WATER, SANITATION AND HYGIENE IN LATIN AMERICA AND THE CARIBBEAN*, https://www.google.com/url?sa=t&rct=j&q=&esrc=s&source=web&cd=&ved=2ahUKEwjmpriy58yLAxWrT6QEHfaIKTYQFnoECBcQAQ&url=https%3A%2F%2Fwashdata.org%2Freport%2Fjmp-2022-regional-snapshot-lac&usg=AOvVaw0AVFcyvD-FufUpOPI3FqeD&opi=89978449. [4]

WHO/UNICEF (2022), *Progress on drinking water, sanitation and hygiene in latin america and the caribbean 2000-2020: 5 years into the sdgs*, https://www.google.com/url?sa=t&rct=j&q=&esrc=s&source=web&cd=&cad=rja&uact=8&ved=2ahUKEwiC9N2uzqOFAxWXcaQEHd6mCdYQFnoECBEQAQ&url=https%3A%2F%2Fwashdata.org%2Freport%2Fjmp-2022-regional-snapshot-lac&usg=AOvVaw0AVFcyvD-FufUpOPI3FqeD&opi=89978449. [12]

World Bank (2019), *Quality Unknown: The Invisible Water Crisis*, https://www.worldbank.org/en/news/feature/2019/08/20/quality-unknown. [21]

World Bank (2016), *The World Bank and the International Water Association to Establish a Partnership to Reduce Water Losses*, https://www.worldbank.org/en/news/press-release/2016/09/01/the-world-bank-and-the-international-water-association-to-establish-a-partnership-to-reduce-water-losses. [30]

World Economic Forum (2020), *WEF Report*. [24]

## Notes

[1] Arup, Ellen MacArthur Foundation, & Antea Group (2019). Water and Circular Economy: A Whitepaper. Available at: https://www.arup.com/-/media/arup/files/publications/w/water_and_circular_economy_whitepaper.pdf; Brears, R. (2019). "Introduction." In Palgrave Studies in Climate Resilient Societies: Developing the Circular Water Economy. Springer International Publishing, Cham. Available at: https://doi.org/10.1007/978-3-030-32575-6_1; Morseletto, P., Mooren, C., & Munaretto, S. (2022). "Circular Economy of Water: Definition, Strategies, and Challenges." Circular Economy and Sustainability, Vol. 2(4), pp. 1463-1477. Available at: https://doi.org/10.1007/s43615-022-00165-x; Sauvé, S., et al. (2021). "Circular Economy of Water: Tackling Quantity, Quality, and Footprint of Water." Environmental Development, Vol. 39, p. 100651. Available at: https://doi.org/10.1016/j.envdev.2021.100651; Smol, M., Adam, C., & Preisner, M. (2020). "Circular Economy Model Framework in the European Water and Wastewater Sector." Journal of Material Cycles and Waste Management, Vol. 22(3), pp. 682-697. Available at: https://doi.org/10.1007/s10163-019-00960-z; World Bank (2021). Water in Circular Economy and Resilience (WICER). Available at: https://www.worldbank.org/en/topic/water/publication/wicer

[2] Latin America comprises Argentina, Belize, Bolivia, Brazil, Chile, Colombia, Costa Rica, Cuba, Dominica, Ecuador, El Salvador, Grenada, Guatemala, Guyana, Haiti, Honduras, Jamaica, Mexico, Nicaragua, Panama, Paraguay, Peru, Puerto Rico, Saint Kitts and Nevis, Saint Lucia, Saint Vincent and the Grenadines, Suriname, Trinidad and Tobago, Uruguay, Venezuela.

[3] SDG 6 on Water and Sanitation seeks to ensure safe drinking water and sanitation for all, focusing on the sustainable management of water resources, wastewater and ecosystems, and acknowledging the importance of an enabling environment.

[4] In the water sector, it is commonly understood as a dimensionless ratio between water used and water withdrawn. In contrast, the agricultural sector often employs WUE to gauge the efficiency of crops in producing biomass and harvestable yield. This ambiguity has led to challenges in standardising the concept.

# 2 The governance of the circular water economy in Latin America

This chapter examines the circular economy policies of the selected Latin American countries and assesses how each dimension of the circular water economy – reduction, reuse, recycling, and recovery – is addressed by sectoral policies in the region. The chapter concludes by identifying governance gaps that need to be addressed to achieve a fully integrated circular water economy in Latin America.

## Circular economy policies in Latin America

In Latin America, national and subnational governments are increasingly developing long term visions towards a circular economy in the form of strategies, roadmaps and plans to foster sustainability and resource efficiency (Box 2.1). Half of these countries – Brazil, Colombia, Costa Rica, Mexico, and Uruguay – have formulated, or are set to implement, national strategies, while Chile, Paraguay, and Peru have opted to develop Circular Economy Roadmaps. Furthermore, Uruguay's earlier Circular Economy Action Plan served as a foundational initiative leading to the formulation of its strategy. In Mexico, a national circular economy policy has been under development since 2023. Some countries have also begun to develop circular economy regulations. For example, Brazil and Costa Rica drafted legislative proposals in 2022 and 2023, respectively. Additionally, Buenos Aires and Mexico City have demonstrated leadership in advancing circular economy initiatives at the local level by enacting subnational circular economy laws in 2021 and 2023.

### Box 2.1. Examples of circular economy policies in Latin America

- In Argentina, while there is no overarching national circular economy policy, various initiatives have emerged at both the national and subnational levels. The **2021 National Strategy for Sustainable Consumption and Production** led by the Ministry of Environment and Sustainable Development acknowledged the circular economy as a means to achieve more sustainable consumption and production patterns, however it does not set specific objectives or targets related to the circular economy (Ministry of Environment and Sustainable Development, 2021[1]). In 2019, the Association for the Study of Solid Waste created a Circular Economy Coalition gathering business, academia and NGOs. In 2019, the Coalition drafted a document outlining preliminary considerations for the future development of a national strategy, including the need for public procurement policies with circular criteria, regulatory updates in the field of waste management, and new regulations for Extended Producer Responsibility (EPR) (ARS, 2019[2]). At the subnational level, the Autonomous City of Buenos Aires enacted the 2021 Framework Law on Circular Economy (Law No. 6468), establishing the principles and guidelines for implementing the circular economy within its territory, and launched in 2024 the 2024-2028 Circular Economy Strategy which focuses on four main areas for intervention: textile, food, waste from vehicles and energy (City of Buenos Aires, 2024[3]).
- Brazil has taken several steps towards a circular economy transition. First, in 2019, the federal government, as part of the National Regional Development Policy, launched a Circular Economy Roadmap, focusing on recycling, renewable energy and water sources (MIDR, 2021[4]). By the end of 2020, the Roadmap led to the creation of two subnational hubs, the Cerrado Circular and Paraíba Circular. Both formed management committees and created project portfolios. Second, in 2022 Brazil started to work on the Circular Economy Bill, which is still under procedure (Federal Senate of Brazil, 2022[5]). Third, in 2024, Brazil enacted the Decree N° 12.082, which establishes the National Strategy for the Circular Economy (Government of Brazil, 2024[6]) and the National Forum for the Circular Economy, a governance body responsible for advice, monitoring and evaluation of the National Strategy.
- In 2021, Chile launched the 2040 Circular Economy Roadmap that targets green jobs creation, reduction of waste generation, enhancement of material productivity and increase in recycling rates, both overall and for municipal solid waste. Additionally, it focuses on recovering sites affected by illegal disposal (Government of Chile, 2021[7]).
- In 2019, Colombia launched the National Circular Economy Strategy, which prioritises material, water and energy efficiency, with actions including the development of monitoring frameworks

and the implementation of measurement standards (Ministry of Environment, 2019[8]). It focuses on six lines of action based on metabolism analyses: i) flow of industrial materials and mass consumption products; ii) flow of packaging materials; iii) flow of biomass; iv) energy sources and flows; v) flow of water; and vi) flow of construction materials. The Strategy presents a series of indicators to measure the progress. At subnational level, seven regional circular economy pacts have been signed to advance in the National Circular Economy Strategy implementation, including the 2018-22 Regional Pact of Bogota and the 2018-22 Regional Pact of Antioquia in Medellin.

- In 2023, Costa Rica introduced both the National Circular Economy Strategy and the Circular Economy Bill (Government of Costa Rica, 2023[9]). The National Circular Economy Strategy establishes objectives, indicators and targets for the materials, waste, energy and water flows, as well as for environmental impact, competitiveness, just transition, economic development, industry and innovation and climate change adaptation. The Circular Economy Bill, still under discussion, addresses the circular economy from a value chain approach, including measures to be taken in five value chains: construction and building, commerce and services, the agricultural sector, manufacturing industry and the water value chain.
- Mexico announced in 2023 the development of the National Circular Economy Strategy and Action Plan by the Department of Environment and Natural Resources (Government of Mexico, 2023[10]). It aims to minimise resource consumption and waste by promoting design, maintenance, repair, and long-lasting recycling, as part of the National Development Plan (2019-2024) and contribute to climate mitigation objectives outlined in the updated 2020 Nationally Determined Contribution. To advance in this effort, in 2024, Mexico published the Diagnosis for the Development of the National Circular Economy Strategy, conducting a national metabolism analysis of the material, energy, and water flows and identifying key sectors for the development of the national strategy (Government of Mexico, 2024[11]). At the subnational level, the 2023 Circular Economy Law in Mexico City champions technology adoption and research support to enhance resource efficiency (CDMX, 2023[12]).
- Paraguay's circular economy efforts primarily target packaging through the 2024 Roadmap for Circular Economy of Packaging that promotes recycling and waste reduction (Ministry of Industry and Commerce, 2024[13]). The Roadmap was launched by the Circular Economy Driving Group, coordinated by the Ministry of Industry and Trade, the Ministry of Environment and Sustainable Development and the Moisés Bertoni Foundation, with the support of the Innovation Lab of the Inter-American Development Bank. This strategy aims to increase the post-consumer packaging recycling rate, boost the use of recycled materials in packaging manufacturing, expand the export of secondary raw materials to more countries, and enhance the recognition of grassroots recyclers in municipal waste management plans.
- Peru developed the Circular Economy Pact in 2021 (MINAM, 2021[14]) and the National Circular Economy Coalition in 2023 (Government of Peru, 2023[15]), which led to the 2030 National Circular Economy Roadmap (MINAM, 2023[16]). The Roadmap establishes 10 targets to be achieved by 2030. These include: increasing the circular economy's contribution to GDP and employment by 1%; lowering the country's ecological footprint; cutting greenhouse gas emissions by 40% compared to the business-as-usual scenario; boosting funding for research, development, and innovation in circular economy initiatives by 50%; labelling 25% of mass-consumption products with information on environmental impacts; incorporating circular economy principles into the development plans of over 70% of Regional Governments; valorising at least 10% of municipal solid waste and 70% of non-municipal waste; sourcing 64% of the national electric grid's energy from renewables; and developing financing mechanisms and capacity-building programmes to promote circular practices across sectors (MINAM, 2023[16]). In addition, the country hosted the First International Forum on the Circular Economy

in 2018 and approved the Roadmap to a Circular Economy in the Industry Sector in February 2020.

- In Uruguay, there have been two main circular economy milestones: the 2019 Circular Economy National Action Plan and the 2024 Circular Economy National Strategy. The Action Plan, led by the National Committee on the Circular Economy, allowed for the articulation of some circular economy initiatives in sectors such as forestry, food, packaging waste, and material recovery. This included, for instance, integration of circular economy criteria for public procurement of food and packaging, industry transition through an awareness programme, and the design of a Technological Centre for Circular Bioeconomy. The National Circular Economy Strategy aims to optimise resource use and promote dematerialisation and is grounded in a metabolism analysis at the national level of five priority flows – biomass, construction, housing and infrastructure, energy and mobility, consumer goods, and water. It is an inter-institutional effort involving four ministries – the Ministry of Environment, the Ministry of Industry, Energy, and Mining, the Ministry of Livestock, Agriculture, and Fisheries, and the Ministry of Economy and Finance – along with actors from the private sector, academia, and civil society organisations.

In recent years, water resources and sanitation have been included in circular economy policies across Latin America. According to the 2020 OECD Survey on the Circular Economy in Cities and Regions[1], 65% of surveyed cities and regions acknowledge water and sanitation as essential component of circular economy strategies (OECD, 2020[17]). Among the 10 surveyed Latin American countries in this report, seven have already implemented circular economy policies, while Mexico is currently in the process of developing one. However, six countries – Brazil, Chile, Colombia, Costa Rica, Peru, and Uruguay – explicitly include water resources in their circular economy policies (Table 2.1). Early initiatives, such as Uruguay's 2019 Circular Economy Action Plan and 2019 Brazil's Circular Economy Roadmap, set the stage for ad hoc measures and localised efforts to include water, like the integration of circular economy criteria for public procurement of food in Uruguay or the development of a digital platform to promote the adoption of water efficient technologies and establishment of circular economy hubs in Brazil. However, more recent policies since 2023, such as Costa Rica's Circular Economy Bill and Strategy and Mexico's Diagnostic for the Development of the National Circular Economy Strategy (Box 2.2), incorporate national-scale metabolic analyses of water, material, and energy flows, alongside the introduction of concrete targets and monitoring indicators. Additionally, countries are starting to develop dedicated policies for the circular economy in the water sector, such as Colombia's 2020 Circular Economy in the Management of Drinking Water Services and Wastewater Handling (CONPES 4004) and Peru's 2024-2030 Roadmap to a Circular Economy in Drinking Water and Sanitation.

**Box 2.2. Leveraging metabolism analysis to shape a circular economy strategy in Mexico**

In 2024, Mexico published its Diagnostic for the Development of the National Circular Economy Strategy. To lay the groundwork for this strategy, which had been under development since 2023, Mexico has undertaken a comprehensive metabolic analysis of its economy, assessing the inputs, uses, and outputs of three key resource flows at the national level: 1) materials, including biomass, fossil fuels, metallic and non-metallic minerals, mixed products, and waste streams; 2) water, covering water use by type and sector, such as industrial activities and electricity generation, as well as sources, including surface and groundwater; and 3) energy, examining Mexico's electricity mix, disaggregated by energy source. This metabolic analysis is represented for each resource flow using the **Sankey diagram**, providing a visual of economic flows.

Source: Government of Mexico (2024) Bases para la Elaboración de un Diagnóstico de la Estrategia Nacional de Economía Circular en México, https://www.gob.mx/semarnat/prensa/semarnat-presenta-las-bases-del-diagnostico-para-transitar-a-una-economia-circular

Circular economy initiatives in Brazil, Chile, Colombia, Costa Rica, Peru, and Uruguay share a common objective of integrating water resource management into their broader sustainability strategies, emphasising water reuse, efficiency, and the reduction of environmental impacts. For example, Uruguay's 2024 National Circular Economy Strategy identifies water as a priority resource flow and promotes measures such as recirculation technologies and resource recovery from wastewater. Similarly, Costa Rica's 2023 Circular Economy Bill emphasises rainwater harvesting and wastewater reuse in key sectors, including agriculture and construction. Brazil's Circular Economy Roadmap highlights water reuse as a key component of promoting circularity across various sectors, particularly in the semi-arid regions. Chile, on the other hand, has implemented a more targeted approach through its 2021 Circular Economy Roadmap, which specifically addresses water resources via initiatives to enhance information systems and updates regulations for waste management, particularly greywater reuse. Despite progress in some countries, significant gaps remain, particularly in the systematic integration of measures for the recovery of energy and materials. While Peru and Uruguay address nutrient recovery and biosolid recycling, many countries fail to prioritise recovery or adequately target underrepresented sectors, such as urban water use and energy.

The distinctive features of each country's strategy are shaped by their unique environmental challenges and governance structures. Colombia's approach, through the 2020 National Circular Economy Strategy and CONPES 4004, relies on targets for increasing water productivity and reducing water pollution, supported by detailed monitoring and evaluation frameworks. Costa Rica's strategy focuses instead on integrating water management into broader eco-friendly practices, such as public procurement and sector-specific guidelines. Peru and Uruguay prioritise water efficiency and recycling within the context of their broader sustainability goals, with Uruguay notably identifying water as one of its five priority flows in the 2024 National Circular Economy Strategy.

Despite the distinct challenges each country faces, common patterns emerge in their efforts to advance circular economy goals. Brazil and Colombia face the need for effective implementation of their circular economy strategies, particularly in regions with significant water scarcity and pollution issues. In Chile and Costa Rica there are regulatory hurdles, specifically for adopting new technologies and practices in water reuse and wastewater management. In Peru and Uruguay, the challenge lies in ensuring that newly established roadmaps and strategies can be effectively operationalised in sectors like water and sanitation, where infrastructure and resource limitations are significant. Despite these challenges, all Latin American surveyed countries share a commitment to fostering innovation, enhancing regulatory frameworks, and building partnerships across sectors to achieve their circular economy objectives.

Funding mechanisms to support circular water economy policies show diverse approaches, with some countries leveraging innovative financing tools and others focusing on more traditional public funding. Countries like Colombia, Costa Rica and Uruguay have developed comprehensive financial strategies that integrate green financing, incentives for businesses, and public-private partnerships. For instance, Colombia encourages private-sector participation through favourable credit conditions and seed capital. Costa Rica offers tax credits, accelerated depreciation, and income tax reductions to promote water-saving technologies and innovation. Uruguay's approach includes financial tools with circularity criteria to guide investment. In contrast, Brazil and Paraguay exhibit a more limited approach, with fewer details available on dedicated financing mechanisms. In Peru, emerging financing mechanisms, including between the IDB and the Peruvian Federation of Municipal Savings and Credit Banks, along with green bonds issued by Banco de Crédito del Peru, support circular economy initiatives. In general, the adoption of blended finance models, along with targeted incentives and capacity-building support, is critical for scaling up circular water initiatives. However, the lack of detailed financial strategies in some countries could impede the effective implementation of circular water policies in those countries.

## Table 2.1. National circular water economy policies in surveyed Latin American countries

| Country | Policy | Year | Explicit mention of circular water economy dimensions | | | Water-related objectives, targets and measures | General and water-specific funding mechanisms |
|---|---|---|---|---|---|---|---|
| | | | Reduction | Reuse & recycling | Recovery | | |
| Brazil | Circular Economy Roadmap | 2019 | ✗ | ✓ | ✗ | Promote the adoption of technologies that optimise water resource usage and reuse in semi-arid regions. To this end, it was developed the Sabiá Platform, a web that features a database of available technologies and education courses. | ✗ |
| | Circular Economy Bill | 2022 | ✗ | ✗ | ✗ | Raise public awareness on the better use of resources, including energy, water, and raw materials. | The Just Transition Mechanism provides financial resources to promote investments, access to loans, startup creation, and innovation and helping high-carbon sectors adopt low-carbon technologies. |
| | National Strategy for the Circular Economy | 2024 | ✗ | ✗ | ✗ | ✗ | Financing schemes, incentives for public procurement of circular goods and services and appropriate tax policies to reduce pollution and waste. |
| Chile | Circular Economy Roadmap | 2021 | ✗ | ✓ | ✗ | • Strengthen information systems for environmental impact modelling, including indicators on water usage (e.g. resource usage, carbon footprint, energy usage, material flow analysis, and waste generation).<br>• Update the regulatory framework to establish standards for grey water reuse systems. | It includes actions such as channeling international impact investing funds towards national circular economy projects, defining funding priorities and designing public financing fund programmes to incentivise the development of secondary markets and ensuring that the financial sector expands its offer of green financing for circular projects and firms |
| Colombia | 2018-2022 National Circular Economy Strategy | 2019 | ✓ | ✓ | ✗ | • Increase wastewater recycling projects by 50%.<br>• Rise treated urban wastewater by 54.3%.<br>• Reduce the Loss Index per Billed Subscriber by 20.9%.<br>• Improve the Water Quality Index by 20 points.<br>• Increase water productivity to USD 5 495 by 2030.<br>• Foster collaborative platforms to support circular economy initiatives in the water sector.<br>• Develop monitoring and evaluation frameworks and implement measurement standards in water supply systems. | Creation of incentives for businesses and entrepreneurs (e.g. favorable credit or seed capital) to build capacities, provide access to technical assistance and promote innovation and R&D. |

| | | | | | | | |
|---|---|---|---|---|---|---|---|
| | Circular Economy in Management of Drinking Water Services and Wastewater Handling (CONPES 4004) | 2020 | ✓ | ✓ | ✗ | With baseline 2018, 4 targets by 2030:<br>• Give access to sanitation solutions in urban areas to additional 763 800 people.<br>• Reach 100% access to safe drinking water (baseline 59%).<br>• Reach 68% of treated wastewater (baseline 42.8%).<br>• Reduce water loss by 30% (baseline 41.4%) | Measure to review and improve resource allocation and spending efficiency in the aqueduct and sewerage sector, as well as adjustments to financing instruments related to the drinking water sector (water usage fees) and wastewater management (retributive fees). The policy includes a participatory budget involving various national entities amounting to nearly USD 8 million. |
| Costa Rica | Circular Economy Bill | 2023 | ✓ | ✓ | ✗ | • Integrates a value chain approach, including the water value chain, emphasising the reuse of treated wastewater, rainwater utilisation, and prioritising measures for rainwater retention and soil infiltration.<br>• Create guidelines for public procurement, e.g. preference for eco-certified products or solutions that reduce water footprint such as water-saving technologies or the rainwater harvesting.<br>• Create sector-specific water-saving guidelines in the construction, agriculture and industrial sectors. | Two main funding areas: innovation in the circular economy – including technology transfer, creation of clusters for innovation and cooperation among private sector and academia – and less water-intensive agriculture, promoting rainwater harvesting, efficient irrigation systems and reducing evaporation, erosion, and nutrient loss.<br>Additionally, it sets incentives for individuals or legal entities engaged in waste repair, recovery, treatment and reintegration into production processes. These incentives are: 1) a tax credit for interest paid on loans classified as green by financial institutions, 2) accelerated depreciation of assets for income tax deductions, and 3) 2% reduction in the income tax. |
| | Nacional Circular Economy Strategy | 2023 | ✓ | ✓ | ✗ | • Reduce total consumptive water use across sectors like agriculture, livestock, forestry, fishing, households and manufacturing.<br>• Increase non-consumptive water use for energy generation.<br>• Increase wastewater treated.<br>• Promote the reuse of water across different sectors.<br>• Reduce physical water losses by water distribution operators | Investments in startegic areas such as climate change adaptation, bioeconomy, industrial digitalisation for a circular economy, and green entrepreneurship.<br>Creation of multilateral organisations, funding agencies and multisectoral partnership for co-financing and financial frameworks focused on compensating socio-environmental impacts. |
| Peru | 2030 National Circular Economy Roadmap | 2023 | ✓ | ✓ | ✓ | • Implement collaborative consumption platforms for resources (e.g. vehicles, machinery, storage spaces, distribution systems, energy and water)<br>• Establish priority areas for the circular economy in the water and sanitation sector, including the reduction and recycling of wastewater, recycling of sludge and reduction of greenhouse gas emissions. | Financing strategy that requires an estimation of implementation costs, technical resources and capacities, identification of funding sources, key stakeholders and risks.<br>Financing mechanisms such as debt-for-nature swaps, global funds for sustainable development, public treasury budgets, regional development funds, and sector-specific funds. |

| | | | | | | | |
|---|---|---|---|---|---|---|---|
| | 2030 Roadmap to a Circular Economy in Drinking Water and Sanitation | 2024 | ✗ | ✓ | ✓ | 6 goals for 2030:<br>• Increase the annual budget for circular practices in the water sector by 20%.<br>• Reuse wastewater managed by operators by 20%.<br>• Utilise 15% more biosolids for soil restoration and agriculture.<br>• Enable up to 20 operators to generate electricity using renewable energy sources.<br>• Adopt circular practices in up to 50% of operators.<br>• Implement conservation measures in at least 40 operators. | ✗ |
| Uruguay | Circular Economy National Action Plan | 2019 | ✗ | ✗ | ✗ | ✗ | ✗ |
| | National Circular Economy Strategy | 2024 | ✓ | ✓ | ✓ | Water as one of the five priority flows with 6 measures:<br>• Integrate circularity into existing water management planning instruments.<br>• Optimise and reduce water use across sectors.<br>• Promote the adoption of technologies for water recirculation and efficiency in the agricultural and industrial sectors.<br>• Deploy technologies for measuring and monitoring water consumption in agricultural, industrial, and urban sectors.<br>• Implement wastewater treatment to recover nutrients, heat and materials.<br>• Encourage the reuse of treated water for non-potable purposes, such as irrigation of green spaces. | • Economic and financial instruments to promote circular water management.<br>• Tariff structures that encourage efficient water use, with incentives for implementing conservation measures.<br>• Financial tools with circularity criteria to guide credit and investment, promoting optimal use and resource recovery. |

Note: Among the surveyed countries Argentina, Honduras and Mexico they do not have a circular economy national policy and the Honduras's 2024 Roadmap for Circular Economy of Packaging does not relate to water. A "✓" indicates it is in place while an "✗" implies it is not in place.
Source: OECD/IDB (2024) OECD/IDB Survey on Water and Circular Economy in Selected Countries of Latin America

## Water frameworks including circular dimensions

The results of the OECD/IDB Survey highlight two complementary policy approaches to address the circular water economy. On the one hand, some countries are developing circular economy policies that, in some cases, mention the link to water or establish specific measures and/or targets related to one or more dimensions of the circular water economy. On the other hand, sectoral policies and initiatives, primarily in the water and sanitation and water resource management policies but also in climate change policies, embed considerations that focus on specific dimensions of the circular water economy.

The OECD/IDB Survey also reveal that the surveyed countries have established goals and principles related to water reduction and efficiency in their regulatory frameworks, accompanied by the development of standards, certifications, and indicators for evaluation and monitoring. In addition, many countries have implemented economic and regulatory instruments as well as water-saving measures, technologies and infrastructure modernisation. Finally, improved data and information on water use and the development of capacities and awareness initiatives are also being developed in most countries. Regulations have been established to delineate the types of water uses and several countries have conducted studies and assessments on the potential uses of wastewater. To a lesser extent, some practices encourage the generation of data and information on water reuse and recycling and engage stakeholders in defining urban reuse plans. However, nutrient recovery and energy generation are not widely considered in current legal and regulatory frameworks in the surveyed countries in Latin America.

The next section provides a detailed review of the relevant regulatory and policy initiatives that have been undertaken in the surveyed countries to help support the circular water economy.

### *Reducing water use and using it efficiently*

All surveyed countries have established regulatory frameworks to promote water reduction and efficiency. They have set targets and implemented specific measures to control water losses, encourage recycling, and optimise usage, with notable progress in laws and reforms across the region.

- In Argentina, the Environmental Management of Waters Regime (Law No. 25688) sets minimum environmental budgets for the preservation and rational use of water.
- In Brazil, Law 11.445/2007 (modified by Law 14.026/2020) establishes the "reduction and control of water losses, promotion of energy efficiency, recycling of sanitary effluents, and rainwater harvesting." The National Water and Sanitation Agency sets standards for the reduction and control of water losses. Ordinance No. 490, 2021 proposes a substantial 90% reduction in losses between 2025 and 2026.
- In Chile, the reform of the Water Code (Law 21.435 of 2023), initially enacted in 1981, promotes a balance between efficiency and security in the productive uses of water. The Draft Law 1317909 of 2020 encourages the rational use of water in the productive sector.
- In Colombia, Law 373 of 1997 defines annual targets for reducing losses in aqueduct systems and integrates water-efficient programmes into regional and municipal plans. Moreover, the 2010-2022 National Policy for the Integrated Management of Water Resources establishes the principle of "savings and efficient use" of water.
- The Water Law in Costa Rica, amended in 2012 addresses water efficiency, sets measures to tackle scarcity and promotes rationalisation.
- In Honduras, the Drinking Water and Sanitation Sector Law identifies water wastage in unauthorised activities as infraction, penalising behaviours that undermine authorised water use. It also mandates service providers to inform users about the benefits of low-consumption sanitary fixtures.

- In Mexico, the General Law of Ecological Balance and Environmental Protection, enacted in 1988 and updated in 2024, mandates authorities to promote water conservation. The CONAGUA promotes efficient water use and manages consumption patterns.
- In Paraguay, Law 1614 of 2000 establishes the regulatory framework for the reduction of water losses, amongst others. The Water Resources Law 3239 of 2007 and the 2023 National Drinking Water Supply and Sanitation Plan promote a rational and efficient use of water.
- In Peru, the Water Resources Law establishes water efficiency as a fundamental principle, and the Framework Law for the Management and Provision of Sanitation Services prioritise efficiency in service provision.
- In Uruguay, the Water Code and Decree 253/979 emphasise the state's and operators' responsibility to promote water conservation and efficient use.

Half of the surveyed countries employ economic and financial instruments to incentivise water use reduction, utilising measures such as fines, usage fees, and dedicated funds to promote efficient water management across various sectors. In Brazil, Decree nº 11,599/2023 allocates financial support to reduce losses in treated water, with goals to be set in the planning instruments by the municipalities. Costa Rica utilises the Fee for Water Utilisation (Executive Decree No. 32868-MINAE), to incentivise improved water resource management across different sectors, including human consumption, agriculture, and industries. Honduras addresses water wastage through the Drinking Water and Sanitation Sector Framework Law (No. 118-2003), imposing fines on users who commit infringements. In Peru, Decree Law 1284 establishes a Safe Water Investment Fund, providing financial support for programmes and projects enhancing the economic and operational efficiency of sanitation service providers. In Uruguay, the National Water Policy advocates for the rational use of water through cultural and educational aspects and introduces charges for usage as an economic instrument to encourage efficient water utilisation.

Most countries have implemented measures to modernise infrastructure and adopt new technologies aimed at improving water efficiency, with initiatives ranging from energy-efficient projects to the adoption of digital tools and low-consumption technologies. In Brazil, the Energy Efficiency Project in Water Supply and the pilot project "Digital Platform Sabiá" implement energy-efficient measures and leverage new technologies for optimising water use. In Colombia, the National Policy for the Integrated Management of Water Resources and its Action Plan encourage the adoption of technologies promoting water savings, alongside improvement of obsolete water supply infrastructure. Also, Law 373 of 1997 and the 2022-2026 National Development Plan further advocate for the adoption of low water consumption technologies, digitalisation, smart metering, and innovation in the water and sanitation sector. Costa Rica, through the Decree No. 30413 and the Code of Hydraulic and Sanitary Installations in Buildings, sets goals for infrastructure expansion and improvements, including in water purification and treatment plants. Honduras, under the General Water Law (No. 181-2009), addresses infrastructure by creating incentives for the implementation of technologies that reduce water use and consumption. In Mexico, the General Law of Ecological Balance and Environmental Protection assigns authorities the responsibility of promoting water-saving measures. In Peru, initiatives under the National Water Resources Policy and the National Water Resources Plan focus on modernising distribution and irrigation systems, establishing efficiency parameters, and promoting research and application of technologies for efficient water use. In Uruguay, the National Water Plan (Decree 205/2017) and the Integrated Waste Management Law incentivise the use of efficient technologies for inputs, including water and energy. Uruguay's National Water Supply and Sanitation Company adopted water loss reduction techniques within two water treatment plants between 2013 and 2019, resulting in a cumulative savings of 89.3 million cubic metres.

Data-driven approaches and capacity-building initiatives are less commonly used for reducing water use and enhancing resource management in the region. In Colombia, Law 373 of 1997 establishes educational campaigns to raise public awareness about the rational and efficient use of water resources. Additionally, Colombia's *Guide to saving and efficient use of water* provides specific directives to environmental authorities and water concessionaires for planning and implementing measures that enhance efficient water use. This initiative focuses on capacity building, ensuring that relevant stakeholders have the necessary knowledge and skills to actively contribute to water conservation efforts. In Costa Rica, the Environmental Accounts of the Central Bank include a water account, providing essential data and metrics to better understand water usage patterns and identify areas for improvement. In 2023, Paraguay introduced the Water Information System, a web platform anticipated to facilitate the generation of efficient and sustainable water resource management (GPSDD, 2023[18]). Similarly, Honduras uses the Agua de Honduras platform to provide reliable and free information on water resources, combining scientific methods and digital technology to support more efficient decision-making in water management. Additionally, the Drinking Water and Sanitation Sector Framework Law (No. 118-2003) seeks to educate users on the advantages of adopting low-consumption habits.

### *Reusing water and recycling treated wastewater*

Several Latin American countries have established regulations to promote the sustainable management of water resources through reuse of untreated water (e.g. rainwater harvesting) and recycling of treated wastewater across various sectors, including domestic, agricultural, and industrial use. In the area of recycling, Colombia's Climate Change Law and Resolution 1256 sets targets for treated wastewater use in domestic applications. Costa Rica further enforces treatment standards and accountability for wastewater management across various sectors. Regarding agricultural use, in Argentina, the Guiding Principles of Water Policy emphasise advocate for water recycling, with regional resolutions focusing on agricultural applications, while Mexico's National Water Law supports municipal wastewater recycling for agriculture, ensuring compliance with health standards. Similarly, Brazil's laws encourage the recycling of sanitary effluents, rainwater harvesting, and greywater reuse in urban and agricultural settings. Peru's National Water Resources Plan and comprehensive legal framework align wastewater recycling with enhancing water resource availability and environmental protection. Industrial wastewater recycling is emphasised Paraguay's 2023 National Plan that supports wastewater technologies in urban and industrial developments, reflecting a similar commitment. Finally, Chile and Honduras focus on incentivising greywater reuse and regulated recycling for ecological preservation and urban applications, with Honduras additionally targeting environmental education and aquifer recharge.

Half of the surveyed countries adopted standards for wastewater recycling. In Brazil, the ANA establishes reference standards for the recycling sanitary effluents, aligning them with environmental and public health standards. Costa Rica, through the Wastewater Discharge and Reuse Regulation (Decree No. 33601), defines maximum permissible limits for the discharge and recycling of wastewater from commercial, industrial, and service activities. In Mexico, NOM-006-CONAGUA-1997 establishes maximum permissible limits for contaminants to ensure the safe recycling of wastewater in public services, contributing to the protection of public health and the environment. In Paraguay, Resolution 222 of 2002 specifies the different types of treated water for domestic supply (special, conventional or simplified water treatment), while Law 5428/2015 defines technical requirements for effluent treatment and specific rules for recycling sewage effluent. Similarly, in Peru, the Supreme Decree No. 003-2010-MINAM approves maximum permissible limits for effluents from domestic or municipal wastewater treatment plants and mandates the establishment of a monitoring programme in these plants, reinforcing the commitment to maintaining water quality standards in recycled wastewater.

A few countries carried out research and assessments to understand opportunities and challenges associated with wastewater recycling. In Argentina, an assessment conducted in 2023 by the Secretary of Infrastructure and Water Policy highlighted the critical need for a National Plan for Wastewater Treatment. This study serves as a foundational step toward recognising the importance of wastewater management and developing strategic plans for its treatment and recycling. In Colombia, the 2018-2030 Water and Basic Sanitation Master Plan undertook a review of regulations to facilitate the recycling of water. Moreover, Colombia's CONPES 3934 Green Growth Policy diagnosed inefficiencies in water resource use and advocated for increased reuse of grey water and rainwater, emphasising the importance of improving efficiency in water use and wastewater treatment. The 2022-2026 National Development Plan in Colombia urged the preparation of a study that analyses and documents techniques for the reuse and recycling of domestic, industrial, agricultural, and rainwater wastewater.

Other practices focus on economic incentives, data and information and stakeholder engagement. In Costa Rica, the 2016–2045 National Policy on Wastewater Sanitation includes two strategic actions: first, it aims to establish a national information system for discharges and water recycling, with the goal of achieving full operationalisation by 2025, emphasising the importance of data and information in effective wastewater management. Second, it foresees to develop a strategy for the safe and valued reuse and recycling of wastewater, actively engaging stakeholders in the decision-making process. In Honduras, the General Water Law (No. 181-2009) designates the Water Authority as responsible for creating incentives to encourage the implementation of wastewater recycle systems. This approach recognises the significance of providing incentives and support to promote sustainable wastewater practices. In Peru, the ProInnóvate National Programme for Technological Development and Innovation, established by Supreme Decree No. 009-2021-PRODUCE, focuses on enhancing the capacity of municipal officials with a specific emphasis on recycling wastewater. This programme addresses the need for skill development, technology adoption, and funding to support effective wastewater recycling initiatives.

### *Recovering energy and materials from wastewater treatment*

Half of the surveyed countries implement standards and regulations to ensure the sustainable management of by-products like sludges and biosolids from wastewater treatment plants. These measures encompass criteria for their use, disposal, or elimination and the creation of databases to track these by-products. Argentina, through Resolution 410/2018, established national criteria for managing sludges and biosolids, allowing their use in afforestation, floriculture, landscape restoration, and more. The resolution also urges the creation of a database for these by-products. Chile set conditions under which sludge can be recovered and reused as compost in Decree 4 of 2009. Colombia's Decree 1287 of 2014 sets criteria for using biosolids, while the National Development Plan 2022-2026 encourages alternatives for by-product recycling. In Costa Rica, the Integrated Waste Management Law and its Regulation address waste from water management, promoting reduction, reuse and recycling, and valorisation. Costa Rica's Regulation for the Management and Final Disposition of Sludge and Biosolids focuses on responsible waste management and reducing the environmental footprint. In Paraguay, Law 5428 of 2015 establishes specific regulations or standards for the recycling of sewage effluent. Peru's Supreme Decree 015-2017-HOUSING regulates the reuse of sludge as biosolids in various activities, promoting their use in agriculture, industry, forestry and ceramics.

Countries are taking steps towards a comprehensive approach to energy generation and decarbonisation in the water and sanitation sector. In Chile, the water utility Aguas Andinas introduced bio-factories as a circular economy solution in 2017, transforming wastewater into electricity, natural gas, and fertiliser for agricultural purposes (Aguas Andinas, 2022[19]). Moreover, the 2030 Sanitation Agenda aims to reduce the carbon footprint of the sanitation sector and promote the circular economy in water resource management by 2030 (Government of Chile, 2023[20]). In Colombia, Law 2169 of 2021, the Climate Change Law, focuses on reducing greenhouse gas emissions by expanding the coverage of domestic wastewater management and utilising biogas in Wastewater Treatment Plants with anaerobic technologies. The 2022-2026 National Development Plan addresses the treatment of industrial wastewater to prevent methane emissions and meet the energy demand of wastewater treatment plants. Costa Rica's National Decarbonisation Plan from 2018 to 2050 underscores the importance of effective wastewater and liquid waste management and reuse as crucial components in driving emissions reduction efforts. In Paraguay, Law 5428 of 2015 establishes specific standards in the use of sewage for the generation of non-conventional energy under the responsibility of the Ministry of Environment and Sustainable Development.

## Governance gaps

According to the OECD/IDB Survey, the three most prominent challenges to transition from a linear to a circular water economy (Figure 2.1) are regulatory gaps (50% of the surveyed countries), insufficient financial resources (40%), and institutional fragmentation (40%) (OECD, 2023[21]).

Figure 2.1. Challenges to the circular water economy in surveyed Latin American countries

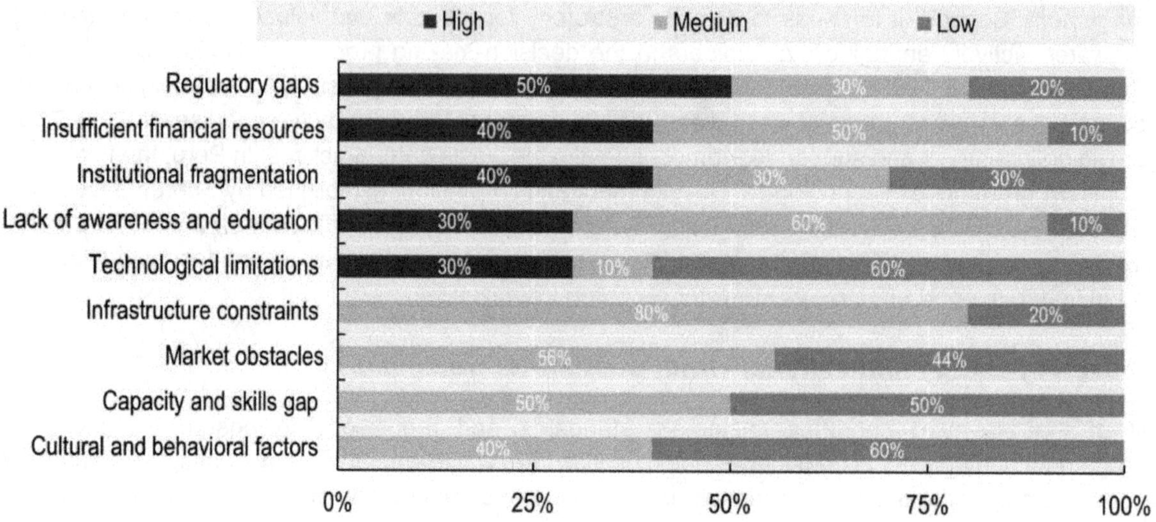


Note: Countries ranked the 9 obstacles in the figure from most to least important, using a scale of 1 to 9.
*Source*: OECD/IDB (2024) OECD/IDB Survey on Water and Circular Economy in Selected Countries of Latin America

- **Regulatory gaps**. A lack of clear, supportive regulations can impede progress towards the circular water economy as existing regulatory frameworks may fall short in addressing the nuances of circular economy practices. Four countries – Argentina, Brazil, Costa Rica, and Mexico – reported that regulatory and policy gaps are the top challenge in implementing a circular water economy. Only half of the surveyed countries, including Argentina, Colombia, Costa Rica, Paraguay and Peru, implement standards and regulations to ensure the sustainable management of by-products like sludge and biosolids from wastewater treatment plants. On the other hand, four countries – Chile, Colombia, Costa Rica and Paraguay – are taking steps to a comprehensive approach towards energy generation and decarbonisation in the water and sanitation sector. Conflicting regulations may create additional roadblocks, hindering the seamless implementation of circular water initiatives. **For example, in Peru, while the water and sanitation regulator SUNASS operates under a strategic plan that aims to optimise water usage and promote sustainability, significant challenges remain in translating these ambitions into practice. The lack of technical capacity, limited enforcement mechanisms, and insufficient integration with broader policies have hindered tangible progress in SUNASS prioritising circular economy principles within water and sanitation services.**
- **Insufficient funding and financial incentives**. The successful transition to circular water management practices requires substantial upfront investments in new technologies, infrastructure upgrades, and comprehensive training programmes, amongst others. The lack of robust financial incentives or support from governmental and institutional bodies represent a barrier to the widespread adoption of circular economy initiatives in the water sector. Three of the ten surveyed countries – Chile, Colombia and Uruguay – reported the lack of financing or financial incentives as the main challenge in transitioning to a circular economy in the water sector. While half of the

countries, namely Brazil, Costa Rica, Honduras, Peru and Uruguay, have economic incentives to encourage the reduction of water use, economic instruments are still needed to encourage the recycling of treated wastewater and the recirculation of by-products. Honduras stands out as the only country whose General Water Law promotes the creation of incentives to encourage the implementation of wastewater recycling systems.

- **Institutional fragmentation**. Overcoming hurdles in circular water management requires a holistic vision, currently jeopardised by institutional fragmentation within the water sector in the surveyed countries. A circular approach to water requires alignment with energy, waste, agriculture, and urban planning policies, among others. This requires establishing coordination mechanisms between departments and levels of government. None of the countries analysed have specific coordination mechanisms to advance the implementation of circular economy policies in the water sector. This is identified as the main challenge to advance in circularity in the water sector in Paraguay and Peru.
- **Lack of awareness, education and data.** A critical impediment to the widespread adoption of circular water management practices is the lack of awareness among key stakeholders, including industry and civil society. Among all the countries surveyed, Peru and Uruguay, through the ProInnóvate initiative and the National Circular Economy Strategy, respectively, stand out as the only countries that have developed actions for capacity building in the circular water economy. Addressing all dimensions of the circular water economy through public awareness actions and capacity building among practitioners and policy making is a key step in the transition. Civil society may not be familiar with the potential benefits and effective approaches associated with circular water management. Honduras identifies the lack of awareness as the main challenge towards a circular water economy in the country. Data collection and monitoring also remain insufficient in surveyed countries, particularly at the subnational level, further limiting effective implementation. For example, in Colombia, the Drinking Water and Sanitation Regulation Commission introduced performance indicators such as the Standard Billed Subscriber Loss Index[1] which measures commercial and technical water losses and sets regulatory targets. However, compliance assessment reveals significant data gaps, with approximately 55% of large providers and 85% of medium-sized providers lacking sufficient data to evaluate compliance.

Building on the OECD Principles on Water Governance (Box 2.3), enhancing policy coherence and establishing clear roles and responsibilities emerged as the top governance dimensions to address in the context of circular water economy, highlighted by 9 and 8 out of 10 of the countries, respectively (Figure 2.2). Additionally, 7 out of 10 of surveyed countries highlighted regulatory frameworks and financing as relevant governance aspects.

**Figure 2.2. Priority governance dimensions for the circular water economy in surveyed Latin American countries**

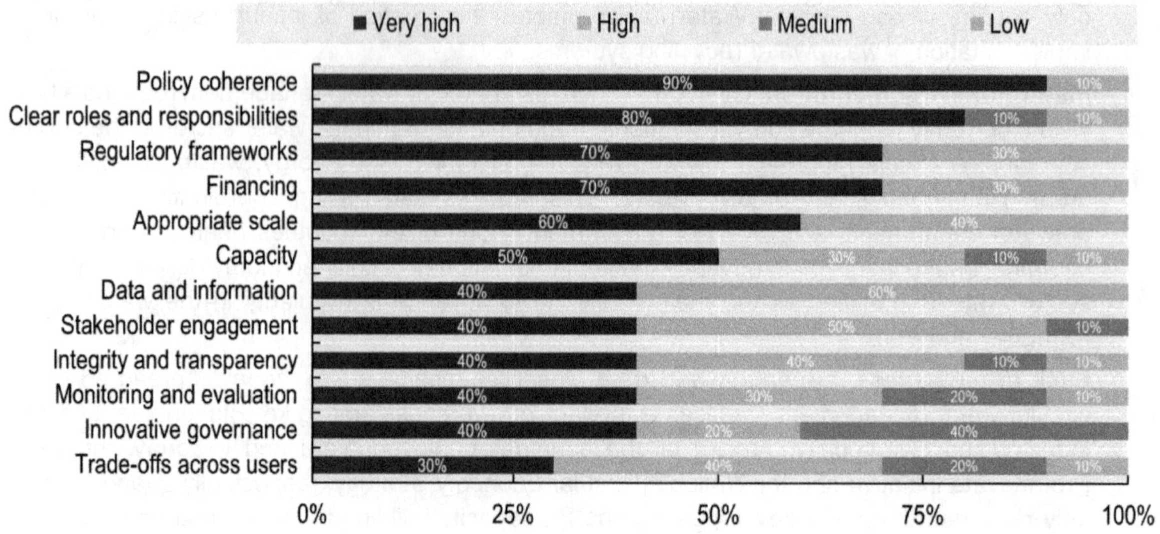


Note: Countries assessed the governance dimensions on a scale of 1 to 4, where 4 denoted the highest level of relevance.
Source: OECD/IDB (2024) OECD/IDB Survey on Water and Circular Economy in Selected Countries of Latin America

## Box 2.3. The OECD Principles on Water Governance

The OECD Principles on Water Governance (Figure 2.3) aim to enhance water governance systems to manage "too much, too little and too polluted" water and foster universal access to drinking water and sanitation in a sustainable, integrated and inclusive way, at an acceptable cost and in a reasonable timeframe. They support effective, efficient and inclusive water governance systems as follows:

**Figure 2.3. OECD Principles on Water Governance**

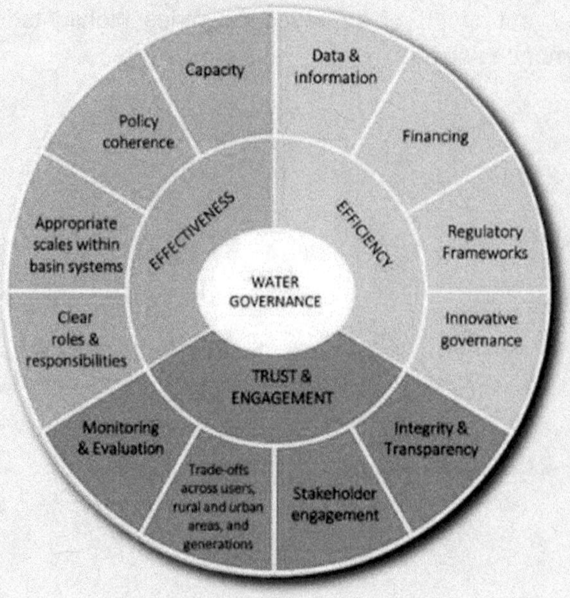

Governance is **effective** when it sets clear policy goals and targets, ensures their implementation, and achieves expected outcomes.

- **Principle 1.** Clearly allocate and distinguish roles and responsibilities for water policymaking, policy implementation, operational management and regulation, and foster co-ordination across these responsible authorities.
- **Principle 2.** Manage water at the appropriate scale(s) within integrated basin governance systems to reflect local conditions, and foster co-ordination between the different scales.
- **Principle 3.** Encourage policy coherence through effective cross-sectoral co-ordination, especially between policies for water and the environment, health, energy, agriculture, industry, spatial planning and land use.
- **Principle 4.** Adapt the level of capacity of responsible authorities to the complexity of water challenges to be met, and to the set of competencies required to carry out their duties.

**Efficiency** involves maximising benefits while minimising costs to society.

- **Principle 5.** Produce, update, and share timely, consistent, comparable and policy-relevant water and water-related data and information, and use it to guide, assess and improve water policy.
- **Principle 6.** Ensure that governance arrangements help mobilise water finance and allocate financial resources in an efficient, transparent and timely manner
- **Principle 7.** Ensure that sound water management regulatory frameworks are effectively implemented and enforced in pursuit of the public interest.
- **Principle 8.** Promote the adoption and implementation of innovative water governance practices across responsible authorities, levels of government and relevant stakeholders.

**Trust and engagement** focus on building public confidence and inclusiveness through democratic legitimacy and fairness.

- **Principle 9.** Mainstream integrity and transparency practices across water policies, water institutions and water governance frameworks for greater accountability and trust in decision-making.
- **Principle 10.** Promote stakeholder engagement for informed and outcome-oriented contributions to water policy design and implementation.
- **Principle 11.** Encourage water governance frameworks that help manage trade-offs across water users, rural and urban areas, and generations.
- **Principle 12.** Promote regular monitoring and evaluation of water policy and governance where appropriate, share the results with the public and make adjustments when needed.

Source: OECD (2015) OECD Principles on Water Governance, http://www.oecd.org/cfe/regional-policy/OECD-Principles-on-WaterGovernance.pdf

## References


Aguas Andinas (2022), *Report Integrado 2022*, https://www.aguasandinasinversionistas.cl/~/media/Files/A/Aguas-IR-v2/2023/reporte/Reporte%20Integrado%20Aguas%20Andinas%202022%201.pdf. [19]

ARS (2019), *Estrategia Nacional de Eocnomía Circular*, http://ars.org.ar/destacados/estrategia-ancional-de-economia-circular/ (accessed on 18 December 2023). [2]

CDMX (2023), *Ley de Economía Circular de la Ciudad de Mexico*, https://plazapublica.cdmx.gob.mx/uploads/decidim/attachment/file/698/Ley_de_Econom%C3%ADa_Circular_de_la_Ciudad_de_M%C3%A9xico.pdf. [12]

City of Buenos Aires (2024), *2024-2028 Primera Estrategia de Economía Circular de la Ciudad de Buenos Aires*, https://buenosaires.gob.ar/sites/default/files/2024-08/Estrategia%20Economia%20Circular%202024-2028.pdf. [3]

Federal Senate of Brazil (2022), *Projeto de Lei n° 1874, de 2022: Institui a Política Nacional de Economia Circular.*, https://www25.senado.leg.br/web/atividade/materias/-/materia/153918. [5]

Government of Brazil (2024), *Estratégia Nacional de Economia Circular*, https://www.planalto.gov.br/ccivil_03/_ato2023-2026/2024/decreto/d12082.htm. [6]

Government of Chile (2023), *Agenda Sanitario 2030*, http://www.sectorsanitario2030.cl/. [20]

Government of Chile (2021), *ROADMAP FOR A CIRCLAR CHILE BY 2040*, https://economiacircular.mma.gob.cl/wp-content/uploads/2022/01/HOJA-DE-RUTA-PARA-UN-CHILE-CIRCULAR-AL-2040-EN.pdf. [7]

Government of Costa Rica (2023), *Proyecto de Ley de la Economía Circular en Costa Rica*, https://d1qqtien6gys07.cloudfront.net/wp-content/uploads/2023/07/23847.pdf (accessed on 8 November 2023). [9]

Government of Mexico (2024), *Bases para la Elaboración de un Diagnóstico de la Estrategia Nacional de Economía Circular en México*, https://www.gob.mx/semarnat/prensa/semarnat-presenta-las-bases-del-diagnostico-para-transitar-a-una-economia-circular. [11]

Government of Mexico (2023), *Trabaja Semarnat en Estrategia Nacional de Economía Circular*, https://www.gob.mx/semarnat/prensa/trabaja-semarnat-en-estrategia-nacional-de-economia-circular?idiom=es. [10]

Government of Peru (2023), *Ministerio del Ambiente lanza oficialmente la Coalición Nacional de Economía Circular*, https://www.gob.pe/institucion/minam/noticias/813231-ministerio-del-ambiente-lanza-oficialmente-la-coalicion-nacional-de-economia-circular. [15]

GPSDD (2023), *Towards a Water Information System in Paraguay*, https://www.data4sdgs.org/sites/default/files/file_uploads/Towards%20a%20Water%20Information%20System%20in%20Paraguay.pdf. [18]

MIDR (2021), *Rota da Economia Circular*, https://www.gov.br/mdr/pt-br/assuntos/desenvolvimento-regional/rotas-de-integracao-nacional/rota-da-economia-circular. [4]

MINAM (2023), *Proyecto de Hoja de Ruta de Economía Circular al 2030*, https://www.gob.pe/institucion/minam/normas-legales/4880223-351-2023-minam (accessed on 1 December 2023). [16]

MINAM (2021), *Conoce los compromisos del Pacto Peruano por una Economía Circular*, https://www.gob.pe/institucion/minam/noticias/509846-conoce-los-compromisos-del-pacto-peruano-por-una-economia-circular (accessed on 1 December 2023). [14]

Ministry of Environment (2019), *Estrategia Nacional de Economía Circular*, https://www.andi.com.co/Uploads/Estrategia%20Nacional%20de%20EconA%CC%83%C2%B3mia%20Circular-2019%20Final.pdf_637176135049017259.pdf. [8]

Ministry of Environment and Sustainable Development (2021), *Estrategia Nacional de Consumo y Producción Sostenibles*, https://www.argentina.gob.ar/sites/default/files/encps_1.pdf. [1]

Ministry of Industry and Commerce (2024), *Hoja de Ruta de la Economía Circular de Envases*, https://www.mic.gov.py/hoja-de-ruta-de-la-economia-circular-de-envases-por-un-nuevo-modelo-de-produccion-y-consumo/. [13]

OECD (2023), *OECD Survey on Water and Circular Economy in Selected Countries of Latin America and the Caribbean*. [21]

OECD (2020), *The Circular Economy in Cities and Regions: Synthesis Report*, OECD Urban Studies, OECD Publishing, Paris, https://doi.org/10.1787/10ac6ae4-en. [17]


## Notes

[1] The survey gathered 51 responses from cities and regions across Europe (38), North and South America (10), Oceania (2), and Asia (1).

[2] Standard Billed Subscriber Loss Index is the Standard Billed Subscriber Loss Index (expressed in $m^3$/subscriber/month) corresponding to the average volume of water per subscriber that is lost in the system, whether through technical or commercial losses. The regulatory standard for this indicator is 6 $m^3$/subscriber/month.

# 3 Roadmap to transition to the Circular Water Economy in Latin America

This chapter presents a Roadmap to transition to the Circular Water Economy in Latin America, addressing governance gaps identified through the OECD/IDB Survey analysis in Chapter 2. The roadmap focuses on three key governance areas: Regulatory and Policy Frameworks, to establish clear and supportive legal structures; Finance and Economic Tools for innovative business models, to allocate funds more efficiently, minimise needs and generate revenues; and Stakeholder Engagement and Capacity Building, to ensure public awareness and acceptance, while equipping professionals with the necessary knowledge while strengthening transparency. Within these key areas, the Roadmap proposes 13 concrete actions to enhance water efficiency, promote resource recovery, and build resilience, offering a practical pathway to circular water management in the region.

## Roadmap to transition to the Circular Water Economy in Latin America

This report proposes a Roadmap to transition to the Circular Water Economy in Latin America that addresses governance gaps identified through the results of the OECD/IDB Survey laid out in Chapter 2 (Figure 3.1). The roadmap focuses on three key governance areas: Regulatory and Policy Frameworks, to establish clear and supportive legal structures; Finance and Economic Tools for innovative business models, to allocate funds more efficiently, minimise needs and generate revenues; and Stakeholder Engagement and Capacity Building, to ensure public awareness and acceptance, while equipping professionals with the necessary knowledge and simultaneously strengthening transparency. Within these key areas, the Roadmap proposes 13 concrete actions to enhance water efficiency, promote resource recovery, and build resilience, offering a practical pathway to circular water management in the region.

Figure 3.1. Roadmap to transition to the Circular Water Economy in Latin America

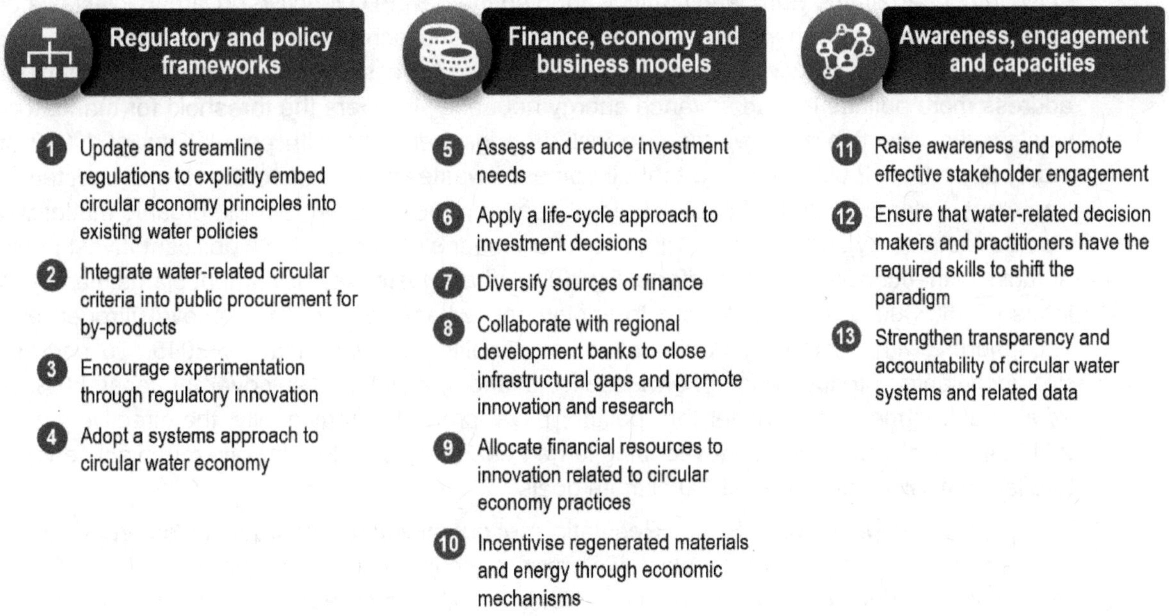


Source: Author's elaboration

## Strengthening regulatory and policy frameworks

### *Action 1: Update and streamline regulations to explicitly embed circular economy principles into existing water policies*

Latin American countries need to identify regulatory gaps and update and streamline their water management regulations to explicitly integrate circular economy principles. Concrete mechanisms to set the regulatory foundations of the circular water economy include establishing clear quality standards for the reuse and recycling of water and materials, creating market structures for the trading of energy and by-products recovered from treatment plants, and incorporating circular economy targets into national and regional water policies. In Europe, a set of regulations and policies, including a dedicated circular economy policy and various water-related policies spanning urban wastewater management, water reuse, sewage sludge management, and renewable energy, exemplifies a holistic approach (Box 3.1).

**Box 3.1. Enabling the circular water economy in Europe: regulatory framework**

The European Commission has defined regulations related to water and wastewater reuse and recovery of materials and energy, emphasising the importance of circular economy principles and climate goals in the water sector. These align with the 2020 European Green Deal aiming for a carbon neutral European Union by 2050. In 2020, the Circular Economy Action Plan, a key component of the European Green Deal's acceleration framework, materialised European efforts to transition to a circular economy. The Plan focuses on sectors that consume the most resources and where circularity potential is high. Specifically, the plan identifies 7 value chains, including "food, water, and nutrients". It requires circular approaches to water reuse in agriculture through the Water Reuse Regulation, the development of an Integrated Nutrient Management Plan, as well as the review of directives on wastewater treatment and sewage sludge. As a result of this plan, the following outcomes were achieved:

- In November 2024, the European Council adopted the new EU Directive on Urban Wastewater Treatment aimed at enhancing water management and achieving a circular economy. The revised directive introduces stricter targets to expand the scope of wastewater treatment, address more pollutants, and advance energy neutrality. It lowers the threshold for mandatory wastewater collection and treatment to include agglomerations with populations of 1 000 or more, down from 2 000, ensuring that all domestic wastewater in these areas is connected to treatment systems by 2035. Secondary treatment to remove biodegradable organic matter will also be required by 2035, with exemptions for member states requiring significant investments or those with recent compliance efforts. By 2039, urban wastewater treatment plants managing loads of 150 000 or more will need to implement tertiary treatment to remove nitrogen and phosphorus, with quaternary treatment for micropollutants mandated by 2045. To address micropollutants, producers of pharmaceuticals and cosmetics will cover at least 80% of additional treatment costs under the “polluter pays” principle. Additionally, the directive sets a 2045 target for energy neutrality, requiring large treatment plants to generate renewable energy to offset emissions and support EU climate goals.
- In 2023, the European Water Reuse Regulation set out minimum water quality requirements for the safe reuse of treated urban wastewaters in agricultural irrigation; harmonised minimum monitoring requirements, notably the frequency of monitoring for each water quality parameter, and validation monitoring requirements; risk management provisions to assess and address potential additional health risks to human and animals; and possible environmental risks; permitting requirements for producing and supplying reclaimed water; transparency, whereby key information about any water reuse project is made available to the public. Although this new resolution was welcomed by Member States, some national regulation still imped the further development of wastewater reuse in some EU countries, with national standards being more stringent and restrictive than EU ones.
- In 2023, an evaluation of the 1986 Sewage Sludge Directive was conducted as part of the European Circular Economy Action Plan, which showed that the levels of heavy metals in sludge have significantly decreased and that a lot of Member States go beyond the Directive standards. Nevertheless, the Directive is still considered relevant, although more could be done regarding the list of contaminants and their risk to the environment and health.
- The European Commission is also developing an Integrated Nutrient Management Action Plan aiming at helping to deliver on the EU Green Deal's targets to reduce nutrient losses by at least 50% and fertiliser use by at least 20% by 2030 in Europe.

Finally, sewage treatment plant, gas and biogas are identified in the revised Renewable Energy Directive 2023/2413 as energy from renewable sources. As such, their production and use are promoted by the new Directive. The revised directive sets an ambitious renewable energy target of at least 42.5% binding at the EU level by 2030, with a goal of reaching 45%. Furthermore, it aims to enhance energy and nutrient recovery in wastewater treatment plants, recognising their significant energy consumption, estimated at 24 747 GWh per year, equivalent to about 0.8% of total EU electricity generation. Notably, small plants account for 42% of energy consumption in wastewater treatment, while large plants utilise 58%. Energy costs can represent a substantial portion, ranging from 25% to 56% of a wastewater treatment plant's operation and maintenance expenses.

Source: European Commission (2022), Nutrients – action plan for better management, https://ec.europa.eu/info/law/better-regulation/have-your-say/initiatives/12899-Nutrients-action-plan-for-better-management/public-consultation_en; European Commission (2023), Renewable Energy Directive, https://energy.ec.europa.eu/topics/renewable-energy/renewable-energy-directive-targets-and-rules/renewable-energy-directive_en; Joint Research Centre (2019), Water Energy Nexus in Europe; Council of the European Union (2024), Urban wastewater: Council adopts new rules for more efficient treatment, https://www.consilium.europa.eu/en/press/press-releases/2024/11/05/urban-wastewater-council-adopts-new-rules-for-more-efficient-treatment/.

### *Action 2: Integrate water-related circular criteria into public procurement for by-products*

Including water reduction, reusing and recycling, as well as energy and materials recovery requirements in public procurement, can help Latin American countries create a market for circular water economy outputs (clean water, energy and by-products), encourage the adoption of circular business models, and promote the incorporation of secondary materials into material flows. For instance, public procurement can incorporate criteria to prioritise compost made from sewage sludge to enrich soil health in public parks, or investing in biogas for powering public transportation fleets, thus reducing GHG emissions. In 2017, the "Circular Public Procurement in the Nordic Countries" report conducted by the Nordic Council of Ministers, explored the potential for promoting the circular economy and fostering new business opportunities through public procurement and investments, with a particular focus on wastewater treatment. In Finland, the region of Helsinki and the cities of Porvoo and Vaasa promote the recovery of materials and energy from the wastewater treatment through public procurement (Box 3.2).

#### Box 3.2. Promoting circular water business models through public procurement in Finland

In 2014, the City of Porvoo initiated a procurement process to improve the recycling and reuse of phosphorus and nitrogen in biowaste and sewage sludge treatment. This was the first time in Finland that nutrient recycling was integrated into a procurement process. The procurement involved both waterworks and biowaste management facilities. Criteria included a minimum of 80% of the nitrogen delivered to the treatment plant to be used as a fertiliser product or industry chemical, and only 20% may end up in the local wastewater treatment plant.

In 2014, the City of Vaasa embarked on a procurement initiative for a fleet of 12 buses designed to operate solely on biogas derived from organic waste and wastewater sludge processed at local treatment facilities. The contract incorporated performance clauses, offering a rebate to the supplier for exceeding estimated annual consumption efficiency and a refund for underperformance, thus incentivising sustained and dependable service. This procurement not only displaces 280 000 litters of diesel annually but also establishes a circular pathway for local waste by-products while facilitating the expansion of infrastructure to accommodate biogas-fuelled vehicles, benefiting an additional 1 000 cars.

In 2016, the Region of Helsinki carried out a procurement process to acquire and pilot test 4–5 new wastewater treatment technologies. The objective was to assess and evaluate innovative methods for treating and utilising digested sewage sludge and other biomass at the waste treatment centre in Espoo, Finland. Via a market inquiry, proposals were sought outlining the end product, market potential, and operational costs, with the flexibility for value chain collaboration among multiple parties. The final procurement was implemented between 2017 and 2018. Additionally, the City of Helsinki currently offers opportunities for its suppliers to develop and pilot products and services in the area of circular economy on the Testbed Helsinki website which is intended for companies and research, development and innovation operators.

Source: Nordic Council of Ministers (2017) Circular Public Procurement in the Nordic Countries, available at https://norden.diva-portal.org/smash/get/diva2:1092366/FULLTEXT01.pdf; European Union (2017) Public procurement for a circular economy. Good practice and guidance; City of Helsinki (n.d.) Testbed Helsinki, available at https://testbed.hel.fi/

### *Action 3: Encourage experimentation through regulatory innovation*

Latin American countries can develop regulatory frameworks that promote innovation within the circular water economy, encouraging controlled and safe experimentation under clear and safe quality parameters, thereby mitigating health and environmental risks. Promoting regulatory innovation is crucial for developing more cost-effective and resource-efficient solutions, which can significantly reduce water use and treatment costs in the long term. Regulatory sandboxes can facilitate the development of experimental pilot projects on a small scale while managing risks effectively. If successful, the pilot projects can later be scaled up with legislative and regulatory reforms that have already been tried and tested (Box 3.3). This can reduce the time and cost of bringing innovative ideas to market while ensuring safeguards to protect consumers are in place to meet water quality and health standards. However, the short-term implementation of these temporary regulatory sandboxes can compromise the impetus to invest due to the lack of sufficient time to recover capital. For example, France established an authorisation procedure in 2022 for new uses of treated wastewater, offering a regulatory sandbox for experimenting with innovative wastewater reuse within a capped time limit of 5 years.

#### Box 3.3. Regulatory sandboxes: how do they work?

Regulatory sandboxes are policy instruments that facilitate small-scale, live testing of innovations in a controlled market-like environment. Sandboxes are typically employed in cases where the emerging technology is potentially disruptive. It allows the testing of innovative technologies and business models that are not fully compliant with current rules and regulations, by providing temporary suspension of certain mandatory provisions or requirements for those who participate in the sandbox. This means that participants are not required to follow all the regulatory requirements that would normally apply outside the sandbox in the regulated market. In return for this dispensation, participants are required to incorporate appropriate safeguards to insulate the market from risk from their innovative business. This gives participants a safe space to experiment without running the risk of being punished for noncompliance while reducing liability concerns among regulators. Regulatory sandboxes tend to be delivered with the strong presence of a regulator, who also provides monitoring and supervision. Another key aspect of regulatory sandboxes is the establishment of feedback mechanisms that allow regulators to gather evidence of potential needs for change in the existing regulatory framework, to facilitate the creation of more products or business models. As such, regulatory sandboxes entail an "interest in regulatory discovery".

Sources: Federal Ministry of Economic Affairs and Energy, Germany (2019), Regulatory sandboxes: experimental areas for new energy technologies, https://www.bmwk-energiewende.de/EWD/Redaktion/EN/Newsletter/2019/07/Meldung/topthema.html; Zetzsche et al. (2017), From FinTech to TechFin: The Regulatory Challenges of Data-Driven Finance, doi: 10.2139/ssrn.2959925.

### *Action 4: Adopt a systems approach to the circular water economy*

Latin American countries have an opportunity to significantly improve the coordination of policies across the environment, water and sanitation, agriculture, energy, urban and regional development, and industry sectors by establishing integrated coordination mechanisms that address common objectives and avoid conflicts. One way to better coordinate policies is by establishing coordination mechanisms linking these policy areas. Looking beyond the water sector, there are several international examples of coordination mechanisms to boost broad circular economy initiatives (e.g. in Melbourne, Australia; Oulu, Finland; and Toronto, Canada) (OECD, 2020[1]). The city of Toronto, for instance, created a Cross-Divisional Circular Economy Working Group comprised of 11 divisions[1], including the Toronto Water as the municipal division in charge on water and sanitation. Its mandate is to provide informed input, ideas, and feedback during the development of the city's circular economy initiatives. Convening these cross-divisional groups helps the city identify sector-related trade-offs as they move forward with circular economy implementation.

## Finance and economic tools for innovative business models

### *Action 5: Assess and reduce investment needs*

Latin American countries need to take a strategic approach to assessing and reducing investment needs in the water sector. This includes prioritising cost-effective measures such as reducing NRW – addressing water leakages, outdated or inaccurate meters, and excessive water pressure – before resorting to costly infrastructure expansion. Many utilities in low- and middle-income countries lose a third of the water they extract, treat, and distribute, resulting in significant financial and operational inefficiencies. By optimising existing water systems and aligning with circular economy principles, countries can achieve better returns on investment than by simply expanding supply. Effective NRW management can help delay or even eliminate the need for expensive projects like desalination plants, ensuring that resources are directed toward sustainable and high-impact investments.

Successful examples demonstrate the financial and operational benefits of such an approach. In New Providence, Bahamas, the Bahamas Water and Sewerage Corporation implemented a Performance-Based Contract[2], leading to significant reductions in NRW and improved financial performance (IDB, 2018[2]). Similarly, Israel has made NRW reduction a core component of its water management strategy, achieving losses below 5% in some areas through rigorous measurement and intervention (EPA, 2023[3]).

### *Action 6: Apply a life-cycle approach to investment decisions*

Implementing a life-cycle approach (LCA) to inform decision-making ensures data-driven investments, offering both immediate and long-term returns and catalysing the transition to the circular water economy. In particular, adopting a LCA approach to wastewater treatment plant investments can significantly reduce operating and maintenance (O&M) costs while unlocking economic opportunities and reducing environmental impacts. This approach entails assessing the long-term benefits, environmental and innovation externalities, and revenue potential of wastewater treatment infrastructure by incorporating circular water economy principles, such as water reuse and by-product recovery. Traditional wastewater treatment plants, designed primarily to meet discharge standards, can evolve into self-sustaining systems by integrating energy recovery, nutrient extraction, and water recycling technologies. By analysing energy potential and payback periods under scenarios like energy self-sufficiency or biogas sales, plants can lower emissions, reduce dependency on subsidies, and create revenue streams. Concrete mechanisms to achieve this include recovering biogas for energy generation, as demonstrated by Anglian Water in England, where sludge treatment centres generate renewable energy for on-site use and export surplus to power grids (Anglian Water, 2023[4]). Similarly, nutrient recovery from sludge for agricultural use adds further value. Internationally, carbon credit schemes such as the Australian Carbon Credit Unit Schemereward wastewater projects that reduce methane emissions or produce renewable energy. These practices illustrate how O&M costs can be offset while promoting environmental sustainability (Box 3.4).

**Box 3.4. Carbon credits from wastewater treatment projects in Australia**

The Australian Carbon Credits Unit (ACCU) Scheme supports projects aimed at either avoiding the release of GHG emissions or capturing and sequestering carbon from the atmosphere. It was established in 2012 under the Carbon Credits (Carbon Farming Initiative) Act 2011 and is overseen by the Clean Energy Regulator, an independent statutory authority responsible for administering the ACCU Scheme, ensuring project compliance, and managing government carbon abatement contracts.

ACCU Scheme methods and projects can involve various approaches, such as adopting new technologies, upgrading equipment, improving land or business practices, and managing vegetation more effectively to store additional carbon.

Projects earn ACCUs based on the amount of carbon dioxide equivalent (tCO2e) emissions they store or avoid. ACCUs can be sold to private sector buyers and government entities, generating income.

Wastewater sector projects are available under the Domestic, Commercial and Industrial Wastewater method (2015) and the Animal Effluent Management method (2019). Eligible activities include installing equipment to capture and use or destroy methane emissions that would otherwise be released to the atmosphere from deep open anaerobic lagoons. Eligible activities also include upgrading biogas from wastewater treatment to produce biomethane. The biomethane can then be used as a natural gas substitute within Australia, contributing to overall carbon abatement efforts, although uptake of this activity has been very limited under the methods to date.

Source: DCCEEW (n.d.) Australian Carbon Credit Unit (ACCU) Scheme, available at: www.dcceew.gov.au/climate-change/emissions-reduction/accu-scheme

### *Action 7: Diversify sources of finance*

Latin American countries need to diversify beyond public sources of finance. In particular, this involves engaging the private sector more actively to access the required economic capital and acquire the technical skills for the transition to the circular water economy. In the LAC region, while the private sector contributed to 78% of total investment in 2019 (below the OECD average of 84%), only 0.2% of private finance in 2018 was directed to water and sanitation infrastructure (OECD et al., 2023[5]). Blended finance can be an effective instrument to mobilise private finance to bridge finance gaps in water and sanitation. However, this requires strengthening transparency and the enabling environment around investment in general and water-related investment in particular (OECD, 2019[6]). According to the Water Integrity Network and the IDB, corruption, mismanagement, and other integrity failures account for up to 26% of wasted investment in the water sector, contributing to the financial shortfall (Water Integrity Network, 2024[7]). Increasing transparency will reduce uncertainty regarding investment risks and opportunities, helping to make water and sanitation investments more appealing to the private sector. Through appropriate contractual agreements or a combination of financial instruments and mechanisms, it becomes feasible to mitigate various risks. Remaining risks can then be shared with the public sector or commercial co-investors or, alternatively, a financier may choose to assume a specific level of risk on their own balance sheet. However, for such an evaluation to be meaningful, the risks linked to an investment must be transparent and quantifiable. Additionally, it is essential to integrate blended finance strategies with initiatives aimed at enhancing the enabling environment. Blended finance alone cannot offset the impact of an unfavourable enabling context, which may include inadequate policy frameworks and institutional structures. Instead, it should be complemented by efforts to establish a stable and supportive broad policy environment. A weak enabling context marked by poorly designed or absent regulations, policy parameters (such as water prices and tariffs), and institutional arrangements can hinder commercial investments.

### *Action 8: Collaborate with regional development banks to close infrastructure gaps and promote innovation and research*

Multilateral development banks (MDBs) have a critical role to play in closing infrastructure gaps and promoting innovation and research. By unlocking regulatory and financial challenges as part of their "country strategies" and setting incentives to strengthen the policy and institutional frameworks, MDBs can help countries foster a more conducive environment for investment choices. Developing policies and regulations that support circular economy initiatives in the water sector, such as through water reuse standards, incentives for sustainable practices, and penalties for pollution is a way forward. Promoting the circular water economy is in line with the IDB's mission to "foster transformative social and economic progress while actively combating climate change" (IDB, 2023[8]). This implies focusing more prominently on projects generating social and environmental impacts across Latin American countries, beyond each country performance. Regional collaboration should also be promoted to share best practices, resources, and technologies for water management.

### *Action 9: Allocate financial resources to innovation related to circular economy practices*

Latin American countries should use a range of financing tools as a means to direct financial resources towards spurring innovation. For instance, in some countries, water regulators have set up specific funds providing financial resources to support innovation, including innovation related to circular water economy practices. This is the case of Ireland (Box 3.5), where the Commission for Regulation of Utilities in its 2014 Water Charges Plan exercised its discretionary power to allow for Irish Water to undertake EUR 2 million in expenditure over its forthcoming review period through an innovation fund[3]. The purpose of this fund was to encourage Irish Water to invest in research and innovation projects to "explore technological advances and other innovations in areas such as effective customer engagement, energy reduction, treatment processes, infrastructure rehabilitation, increased understanding of customer behaviours, climate change adaptation and environmental compliance, which could ultimately benefit customers."[4]

#### Box 3.5. Incentivising innovation through funding: the case of Ireland

Irish Water is required to submit a proposal to the Commission for Regulation of Utilities (CRU) before incurring expenditure, to demonstrate that the proposed investment in research or innovation meets the qualifying criteria. The proposals for innovation fund expenditure must (a) have a reasonable probability of delivering defined and tangible benefits to customers and (b) the expected benefits of the investment must outweigh its costs. Proposals must target at least one of the following objectives: the provision of safe, secure, and reliable water services; increased understanding of customer behaviours and their drivers and effective customer engagement; enhanced energy savings in the provision of water services; achievement of relevant environmental standards and the objectives of the Water Framework Directive; mitigation of negative climate change impacts; provision of water services in an economical and efficient manner; and improved conservation of water resources.

In addition to requiring Irish Water to seek approval for innovation fund expenditures in advance, reporting requirements are also in place to measure the outcomes of these projects and measure them against expected benefits. As of November 2021, CRU had approved 10 projects whose expenditure could be counted against the innovation fund component since its introduction at the beginning of the 2015 regulatory review period.[5]

Source: CRU (2020) Water Services Innovation Fund Annual Report 2020, CRU21121-Water-Services-Innovation-Fund-Annual-Report-2020.pdf (divio-media.com)

### *Action 10: Incentivise regenerated materials and energy through economic mechanisms*

Latin American countries can make greater use of economic instruments to incentivise the recirculation of by-products generated in wastewater treatment (clean water, biocomponents and energy) into the economy, making them cheaper than linear alternatives (e.g. extractive water, chemical fertilisers, and non-renewable energy). Economic instruments such as tax discounts, environmental taxes, or differentiated tariffs can serve as tools for encouraging or discouraging specific market behaviours. For instance, offering tax discounts for industries that use regenerated water or recycled materials in their production processes can stimulate demand for these resources while reducing overall water consumption and waste generation. Similarly, imposing environmental taxes on industries that excessively pollute or generate wastewater can incentivise the adoption of cleaner production methods and encourage investment in water-efficient technologies. Moreover, implementing differentiated tariffs for water usage based on the degree of treatment and reuse can promote efficient resource allocation and encourage industries to invest in advanced wastewater treatment technologies to produce high-quality reclaimed water suitable for various purposes. Economic instruments can also promote the recovery of energy and materials from wastewater treatment. The Italian regulator ARERA has introduced mechanisms to boost energy neutrality in its 2024-2029 tariff methodology by incentivising electricity self-production and encouraging energy savings through sharing factors (Box 3.6). These instruments can be useful for developing markets around recovered by-products. Adjustments in the prices of treated wastewater, recovered biosolids, or generated energy, as well as exemptions on Value Added Tax (VAT), can incentivise businesses and users to reuse recovered materials and energy. Additionally, extended producer responsibility (EPR) schemes can help develop programmes to reuse materials or stimulate innovation.

#### Box 3.6. Incentive mechanisms to boost energy neutrality

In 2018, the Italian regulator ARERA introduced innovative and multi-sector measures aimed at energy efficiency, plastic use reduction, energy and raw material recovery, and wastewater reuse. The measures include the installation of soft starters and inverters, pressure management, and energy recovery devices for improved energy efficiency. Additionally, high-quality fountains were installed, tap water was promoted as an alternative to bottled water to reduce plastic usage, and biogas production from sludge was implemented. Furthermore, mini hydroelectric power plants were installed, and biopolymers, struvite, and phosphorus were recovered from sludge for energy and raw material recovery purposes. Finally, reclaimed water was reused for irrigation or industrial purposes, along with internal plant reuse for technical water in wastewater treatment processes.

These measures are incentivised through a revenue sharing mechanism. Operators are not compelled to implement such measures as they are not considered by ARERA as being part of mandatory water and sanitation services standards.

The environmental and resource costs include both operational and capital cost components and are used to promote sustainability and resilience. More specifically, some operational components of environmental costs are used to reflect and cover expenditure targeted towards remote control operations aimed at reducing or preventing water losses. Since 2016, ARERA has also introduced 6 macro performance indicators with differentiated regulatory targets according to the operator's efficiency. If the operator fails to achieve the expected improvement, and thus lower water losses, it is penalised.

Source: Bardelli, L. (2023). Report on Water Services & Waste Management, https://www.arera.it/fileadmin/allegati/eventi/230412analistiAmbiente.pdf.

## Stakeholder engagement and capacity building

### *Action 11: Raise awareness and promote effective stakeholder engagement*

Latin American countries need to emphasise transparency and information sharing to ensure that communities understand the benefits of the circular water economy and are engaged in policy making and implementation processes. Community concerns increase as the degree and likelihood of personal contact with recycled water rises. For example, use of recycled water for urban or agricultural irrigation has high levels of acceptance (Po et al., 2005[9]), whereas closer contact, including consumption of recycled water, has lower levels of support (Fielding, Dolnicar and Schultz, 2018[10]). Proposals to use recycled water for drinking supply tend to polarise views in communities, meaning that consultation with stakeholders is a key element when developing recycled water schemes. Tailored communication mechanisms should be implemented, encompassing both online and offline channels. This includes communication campaigns involving the creation of dedicated websites and social media platforms, events for the exchange of knowledge and practices, conferences, etc.

### *Action 12: Ensure that water-related decision makers and practitioners have the skills required to shift the paradigm*

Capacity building activities tailored to different stakeholders in both the public and private sectors can enhance the skills and knowledge in the water sector, fostering a deeper understanding of circular water management principles. For instance, capacity-building initiatives for policy makers could focus on providing them with the tools and knowledge to develop and implement policies that promote circular water practices. Practitioners in the water sector, such as engineers and technicians, could benefit from hands-on training programmes specifically designed to enhance their technical expertise, covering practical topics like the design and operation of decentralised wastewater treatment systems.

### *Action 13: Strengthen transparency and accountability of circular water systems and related data*

Latin American countries need to strengthen transparency in water management and treatment to foster public trust and engagement. This requires monitoring circular water economy data, which goes beyond drinking water and sanitation access rates to include indicators related to water reduction, reuse and recycling, and material and energy recovery. This includes data on indicators such as NRW, water savings from the implementation of conservation techniques and technologies, volume of treated wastewater reused for various purposes, volume of recovered sludge used as fertilisers, and energy savings from recovery during treatment, among others. Regular monitoring of these indicators of water management circularity can help ensure efficient fund utilisation and gauge the impact of circular water initiatives. Regulators from various countries have already established key performance indicators targeting water losses, water, and energy efficiency (Box 3.7).

## Box 3.7. Regulatory KPIs dedicated to water loss reduction and energy neutrality

The association of European Water Regulators (WAREG) assesses key performance indicators used by regulators to evaluate the performance of utilities. The reports notes that 18 WAREG members (Albania, Azores, Brussels, Bulgaria, Estonia, Flanders, Georgia, Greece, Hungary, Ireland, Italy, Kosovo, Latvia, Malta, Montenegro, North Macedonia, Portugal and Romania) use KPIs to monitor and assess NRW / water loss (Table 3.1).

The report highlights that NRW is widely used by regulators for monitoring and assessment, whether expressed as a percentage (11 regulators), in volume per km per day (6 regulators), or in litres per connection per day (2 regulators). Real losses are commonly monitored by 5 regulators, while 3 other regulators use an infrastructure leakage index.

### Table 3.1. Regulatory KPIs in WAREG

| COUNTRY | No | KPI NAME | KPI UNIT |
|---|---|---|---|
| Albania | 1 | NRW | % |
| Azores | 13 | NRW | % |
| Brussels | 7 | DW-Loss02: Infrastructure Leakage Index (ILI) | # |
| Brussels | 8 | DW-Loss03: Real losses by connections | 1/ 1000 connect |
| Bulgaria | 6 | PK4a: Water loss | m3/km/d |
| Bulgaria | 7 | PK4b: Water loss | % |
| Estonia | 1 | Water loss | % |
| Flanders | 3 | Lost water/branch / day | Litter |
| Flanders | 4 | Infrastructure Leakage Index (ILI) | Factor |
| Georgia | 9 | Infrastructure leaking index (ILI) | Ratio |
| Greece | 6 | Water Losses | m3 |
| Hungary | 5 | Water loss | m3/km/day |
| Hungary | 6 | NRW | % |
| Ireland | 11 | Leakage | 0 |
| Ireland | 43 | Leakage Reduction | ML/day |
| Italy | 1 | Water losses per km (M1a) | mc/km/day |
| Italy | 2 | Leakage rate (M1b) | % |
| Kosovo | 5 | NRW | % |
| Latvia | 1 | Water loss | % |
| Latvia | 2 | Water loss | m3/km/year |
| Malta | 4 | Estimated Leakage | 1/prop/day |
| Malta | 5 | Estimated Leakage | m3/km/day |
| Malta | 9 | Unaccounted for water (NRW ) | m3/km/day |
| Montenegro | 4 | NRW | % |
| North Macedonia | 3 | NRW | % |
| North Macedonia | 4 | NRW | m3/km/day |
| Portugal | 9 | AA08 - NRW | % |
| Portugal | 16 | AA15ab - Real water losses (Bulk systems and retail systems with service connection density less than 20 service connections per km) | m3/(km. day) |
| Portugal | 17 | AA15b - Real water losses (Retail systems) | 1/ (service connection day) |
| Romania | 8 | NRW | thousand mc |

Additionally, some water regulators have created incentives for energy recovery through benchmarking regulatory indicators (Table 3.2).

## Table 3.2. Energy neutrality regulatory indicators used by a sample of regulators

| Location | Name of regulator | Name of indicator | Unit |
|---|---|---|---|
| Brussels | BRUGEL | On-site energy production in UWWTPs | kWh |
| Hungary | ERRA | Energy production (own energy) | % |
| Portugal | ERSAR | Self-produced energy | % |

Source: WAREG (2023), Key performance indicators framework in WAREG Member Countries, https://www.wareg.org/documents/kpis-report-2023-wareg-pdf/

## References

Anglian Water (2023), *2025 – 2050 Bioresources Strategy*, https://www.anglianwater.co.uk/SysSiteAssets/household/about-us/bioresources-strategy-2025---2050.pdf. [4]

EPA (2023), *From Water Stressed to Water Secure: Lessons from Israel's Water Reuse Approach,*, https://www.epa.gov/system/files/documents/2023-04/Action%2011.1%20Closeout_Draft_041723_508%20Compliant.pdf. [3]

Fielding, K., S. Dolnicar and T. Schultz (2018), "Public acceptance of recycled water", *International Journal of Water Resources Development*, Vol. 35/4, pp. 551-586, https://doi.org/10.1080/07900627.2017.1419125. [10]

IDB (2023), *IDB Group Institutional Strategy: Transforming for Scale and Impact*, https://www.iadb.org/en/who-we-are/institutional-strategy. [8]

IDB (2018), *Case Study: Performance-based Contract for NRW Reduction and Control New Providence, Bahamas*, https://publications.iadb.org/es/publicacion/12918/case-study-performance-based-contract-nrw-reduction-and-control-new-providence. [2]

OECD (2020), *The Circular Economy in Cities and Regions: Synthesis Report*, OECD Urban Studies, OECD Publishing, Paris, https://doi.org/10.1787/10ac6ae4-en. [1]

OECD (2019), *Making Blended Finance Work for Water and Sanitation: Unlocking Commercial Finance for SDG 6*, OECD Studies on Water, OECD Publishing, Paris, https://doi.org/10.1787/5efc8950-en. [6]

OECD et al. (2023), *Latin American Economic Outlook 2023: Investing in Sustainable Development*, OECD Publishing, Paris, https://doi.org/10.1787/8c93ff6e-en. [5]

Po, M. et al. (2005), *Predicting Community Behaviour in Relation to Wastewater Reuse : What drives decisions to accept or reject ? Water for a Healthy Country National Research Flagship*, https://www.researchgate.net/publication/242602980_Predicting_Community_Behaviour_in_Relation_to_Wastewater_Reuse_What_Drives_Decisions_to_Accept_or_Reject. [9]

Water Integrity Network (2024), *Water Integrity Global Outlook 3*, https://www.waterintegritynetwork.net/_files/ugd/3abf8e_7081cf31513f4898bd79656bbb75b4b5.pdf. [7]

## Notes

[1] The 11 divisions are: Solid Waste Management Services (SWMS); Purchasing and Materials Management (PMMD); City Planning; Economic Development and Culture; Environment and Climate Division; Corporate Real Estate Management; Parks, Forestry and Recreation; Transportation Services; Toronto Water; Toronto Public Health; and Engineering and Construction Services.

[2] Performance-based contracts (PBCs) are specialised public-private partnerships increasingly used by utilities to manage NRW. These contracts link payment to contractors' performance, incentivising them to meet NRW reduction targets. PBCs enable utilities to access expertise and equipment while retaining control over operations and assets. IWA (n.d.), Performance-based contracts for non-revenue water management; PPIAF (2023), Using performance-based contracts to reduce non-revenue water; World Bank (2006), The challenge of reducing non-revenue water in developing countries; World Bank (2018), The use of performance-based contracts for non-revenue water reduction.

[3] Commission for Energy Regulation (Ireland), Water Charges Plan Decision Paper, October 8th 2014 (CER/14/746)

[4] Commission for Energy Regulation (Ireland), Water Services Innovation Fund, April 10th 2015 (CER/15/076), p. 2

[5] Commission for Regulation of Utilities (Ireland), Water Services Innovation Fund, Annual Report 2020, November 16th 2021 (CRU/21/121)

# Annex A. Country participants

### Table A A.1. Country participants in the OECD/IDB report on Circular Water Economy in Latin America

| Country | Period | Participants | Position/Institution |
|---|---|---|---|
| Argentina | 2023-24 | Jose Maria Regueira | National Director of Drinking Water and Sanitation, Ministry of Public Works |
| | 2024-25 | Javier Mijangos | National Director of Drinking Water and Sanitation, Ministry of Public Works |
| Brazil | 2023-25 | Tauana Monteiro | Specialist in Public Policy and Government Management, National Water and Sanitation Agency (ANA) |
| Chile | 2023-25 | Carlos Estevez | Water Area Coordinator, Ministry of Public Works |
| Colombia | 2023-25 | Claudia Andrea Ramirez | Director of Urban Development, National Planning Department (DNP) |
| | 2023-24 | Ricardo Alberto Bula Torres | Deputy Technical Director of Water and Sanitation, National Planning Department (DNP) |
| | 2024-25 | Natasha Valentina Garzón Yepes | Deputy Technical Director of Water and Sanitation, National Planning Department (DNP) |
| | 2024-25 | Jefferson David Ladino Malagón | OECD Technical Secretariat, National Planning Department (DNP) |
| | 2024-25 | Daniel Santiago Higuera | Director of Urban Development, National Planning Department (DNP) |
| Costa Rica | 2023-25 | Andrez Lazo | Director, Costa Rican Institute of Aqueducts and Sewers (AyA) |
| Honduras | 2023-25 | Ricardo Panchamé | General Director, National Autonomous Water and Sewerage Service (SANAA) |
| | 2023-25 | Manuel de Jesús Martínez | Deputy General Manager, National Autonomous Water and Sewerage Service (SANAA) |
| | 2023-25 | Joan Zelaya | Sub-Secretary General, National Autonomous Water and Sewerage Service (SANAA) |
| Mexico | 2023-25 | Germán Martínez Santoyo | Director, National Water Commission (CONAGUA) |
| | 2023-25 | Griselda Medina Laguna | Deputy Manager of Management and Evaluation of Projects with External Credit, CONAGUA |
| | 2024-25 | Silvia Pilar Chavez Cereceda | Deputy Manager of Management and Evaluation of Projects with External Credit, CONAGUA |
| Paraguay | 2024-25 | Claudia Crosa | Director of Water and Sanitation, Ministry of Public Works and Communications |
| Peru | 2024-25 | Max Carbajal | Director of Policies and Regulation of Sanitation, Ministry of Housing, Construction, and Sanitation |
| Uruguay | 202325 | Andrea Gamarra | Director of Drinking Water and Sanitation Division, National Directorate of Water (DINAGUA) |

www.ingramcontent.com/pod-product-compliance
Lightning Source LLC
LaVergne TN
LVHW060455061126
851857LV00016B/1839